Walter Harvey Weed

The formation of travertine and siliceous sinter by the

vegetation of hot springs

Walter Harvey Weed

The formation of travertine and siliceous sinter by the vegetation of hot springs

ISBN/EAN: 9783337204532

Printed in Europe, USA, Canada, Australia, Japan

Cover: Foto ©berggeist007 / pixelio.de

More available books at **www.hansebooks.com**

FORMATION OF TRAVERTINE AND SILICEOUS SINTER BY THE VEGETATION OF HOT SPRINGS.

BY

WALTER HARVEY WEED.

CONTENTS.

ILLUSTRATIONS.

617

FORMATION OF TRAVERTINE AND SILICEOUS SINTER BY THE VEGETATION OF HOT SPRINGS.

By WALTER HARVEY WEED.

INTRODUCTION.

Among the many interesting natural phenomena that claim the attention of the visitor to the Yellowstone National Park, the geysers and hot springs rank first in general interest. Their novelty and beauty are sure to attract universal admiration, while the vast quantities of hot water that flow from the ground are convincing evidences of the nearness of internal heat. These steaming fountains and boiling pools are usually surrounded by snowy white borders of mineral matter deposited by the hot waters. At the Mammoth Hot Springs this consists of carbonate of lime, that forms the unique marble terraces and pulpit basins of those springs. (Pl. LXXIX.) At the Geyser basins the waters deposit silica, that forms the fretted rims of the pools and the beautifully beaded and coral-like deposits of the cones, and covers large areas of ground about the springs with a sheet of white and glaring sinter. Not only are the occurrence and the nature of these deposits such as make them of interest to every visitor, but the problem of their origin has proved to be one of the prominent features in the scientific investigation of the hydrothermal phenomena of the park, as it has been found that such deposits are very largely due to the growth and life of a brilliant colored algous vegetation, living in the hot mineral waters.

PLANTS AS ROCK-BUILDERS.

A review of the various geologic agents that have built up the strata forming the earth's crust shows that living organisms have taken an important part in rock formation. The abundance of their remains in ancient as well as the most recently formed sediments shows that the corals and mollusks of all periods have been active rock-builders. The geological work executed by such forms of animal life is therefore quite apparent to the students of nature. On the contrary, the geological work of plant life has not been generally recognized, partly because it is less conspicuous, and partly because the absence of organic remains in many deposits formed in this way has prevented a recognition of the true origin of the rocks.

It has been proved that living plants further geologic change in
several ways; by promoting the disintegration and decay of existing
rocks, by building up new rock formations, and, upon their death, by
starting a series of changes resulting from the action of the decaying vegetable matter upon various mineral substances. New formations are built up by living plants in two ways—by the accumulation
of their plant remains and by the chemical reactions resulting from
the growth and life of the plants: in either case mineral matter is deposited. Where the mineral matter preserves the form and structure of the plant, as is the case with the silica forming the well-known
beds of diatomaceous earth, the origin of the deposit is apparent, but
in many cases no trace of plant structures can be distinguished, even
when thin sections of rocks that are undoubtedly formed by plants
are seen under the microscope. This is true of some marine limestones formed by calcareous algæ, and is especially true of several
classes of deposits heretofore considered to have a purely chemical
origin, such as travertine, siliceous sinter, certain gypsums and
iron ores. In such cases it is only by a careful study of the actual
process of formation of the deposits that we can tell with certainty
their true manner of formation. This has been done in the case of
the deposits formed about the hot springs and geysers of the Yellowstone Park, and it is the purpose of the present paper to show the
origin and manner of formation of these interesting mineral deposits.

VEGETATION OF HOT WATERS.

The presence of organic life in highly heated mineral waters is a
subject of considerable interest not only to students of biology, but
to geological observers as well. It shows the development of life
under very adverse conditions of temperature, and affords an opportunity for the study of the modifying effect of high temperatures
and chemical solutions upon forms found also in ordinary surface
waters. The ability of life to withstand such extreme conditions
shows the possible existence of such forms in the early history of
the earth, when the crust is supposed to have been covered by highly
heated mineralized water. Thus far this subject has received but
little attention, and the data accessible are meager and unsatisfac-
tory, this being especially true of the animal life of hot waters.

The vegetation of hot springs consists entirely of various species
of fresh water algæ, flowerless cryptogamic plants, closely related
to the salt water algæ or sea-weeds. The fresh water species are
less striking and varied than the marine growths, and are generally
composed of green thread-like structures of more or less slimy consistency.[1] It is well known that algæ are abundant in the hot waters
of many and widely separated localities, for, in the various works of

---

[1] Phycology: Prof. Farlow, in Johnson's Encyclopedia.

travel and exploration in which the occurrence of hot springs has been described, mention is frequently made of bright green confervæ living in the hot pools and streams. Where the plants present in thermal waters are of this color their vegetable nature seems to have been readily recognized, but there is reason to believe that the existence of algæ of other colors, such as the red and yellow species common in the Yellowstone springs, has generally been overlooked or the growth mistaken for mineral matter. This is not surprising, as the plants are often incrusted and hidden by mineral material deposited by the hot water, and the organic nature of the substance is often scarcely recognizable even by botanists. Thus in sulphur waters the algæ are very generally incrusted by grains of sulphur, or are inclosed in gypsum, while the vegetation of calcareous springs is often buried in travertine deposited by the water, only the growing tips of the plants being free. Similarly, the threads of algæ living in ferruginous waters are incrusted by oxide of iron, while in siliceous waters such growths are inclosed in gelatinous silica.

In reviewing the literature of this subject, vegetation is found to be a common accompaniment of thermal springs in all parts of the world, but, although the presence of these hot-water growths has been recognized, the conditions under which they exist are rarely given and the plants themselves have been studied and identified at very few localities. Of these the foremost is Carlsbad, Bohemia. Its hot springs have long been noted for their curative properties, and thus they attracted the attention of scientific men at an early date. In 1827, Agardth described the algous growths of these thermal waters,[1] and the botanist Corda[2] figured and described species from these springs in 1835. Schwabe published a paper in 1837[3] in which he describes the occurrence of the algæ, giving the temperatures at which the different species were found, besides figuring and describing the plants themselves. The most important paper, from a geological stand-point, is, however, that published by Prof. Ferd. Cohn in 1862,[4] in which the physiological action of the plant life is shown to cause the deposition of travertine by the hot waters.

Algæ from the hot springs of Italy were described by Meneghiné[5] in 1842, and Ehrenberg says[6] that algæ occur in the hot springs of Ischia at 174° F. to 185° F. Hoppe Seyler[7] found similar vegetation in the hot waters of Lipari at 127° F. The writings of Kützing mention a number of species from European hot springs, and other localities are given by Rabenhart.[8]

The hot springs and geysers of Iceland have been famous for many centuries, but a careful examination of the writings of the

[1] Flora, 1827.
[2] Almanach de Carlsbad, 1835-'36.
[3] Linnæa, 1837.
[4] Abhandl. Schles. Gesell. Naturwiss., Heft 2, 1862.
[5] Monographia Nostochinarum Italicarum: Turin, 1842.
[6] Sachs in Flora, 1864.
[7] Pflugers Archiv, 1875.
[8] Flora Aquæ Dulcis.

numerous travelers who have visited and described them shows that only three authors have mentioned the presence of algous vegetation in the hot waters. Sir William Hooker,[1] who visited Iceland in 1809, found confervæ at the borders of many of the hot springs, where the plants were exposed to the steam and heat of the boiling water. *Confervæ limosa* Dillw. was found in abundance, forming large dark-green patches attached to a coarse white clay, from which it could be easily peeled off. A brick-red confervæ, an *Oscillatoria*, occurred in a similar way, forming large patches several inches square. *Confervæ flavescens* Roth, and a species allied to *C. rivularis*, were abundant in water of a very great degree of heat.

Baring-Gould, who visited the geysers in 1864, found a crimson algæ growing in the spray and overflow of the spring Tunguhver.[2] He collected specimens, which were examined by Rev. J. M. Berkley, who referred them to the genus *Hypheothrix*, common in hot waters all over the world. Lauder Lindsay found two kinds of confervæ in the springs of Langarnes, Iceland, in water so hot that an egg was boiled in it in four to five minutes.[3]

In New Zealand the presence of algæ in the hot springs on the south shore of Lake Taupo was first noted by Hochstetter, who says' the dark emerald-green growth covered the ground where the warm water flowed. The specimens collected by him are described in the Botany of the Novara Expedition.

Algæ from these springs are also described by W. I. Spenser, the highest temperature of the water in which they were found being 136° F. Hochstetter says the temperature of the springs varies between 125° F. and 153° F.

Dr. S. Berggren, of Lund, Sweden, visited the hot spring district of New Zealand in 1874, and collected an extensive series of specimens of the algæ of the region. He states[6] that the algæ, especially *Phycochromaceæ*, but likewise *Confervaceæ* and *Zygnemaceæ*, are to be found growing in great abundance in the rivulets from the hot springs.

These specimens have been studied by Dr. Otto Nordstedt, whose determinations show that the species are chiefly those common in hot waters in other parts of the world, and that several species occur both in hot and cold waters.

Thick masses of slimy confervoid plants line the bottom of a large pool, Tapui Te Koutu, at Rotorua, New Zealand, where the usual temperature is 90° to 100° F., but is 180° with a north or east wind.[7]

[1] Journal of a Tour in Iceland, vol. 1, p. 160.
[2] Iceland : Its Scenes and Sagas.
[3] Bot. Zeitung, 1861, p. 359.
[4] Reise der Oe. Frégatta Novara : Geol., vol. 1, pt. 1, p. 126.
[5] Trans. New Zealand Inst., vol. 15, p. 302.
[6] Kongl. Sv. Vet. Akadémiens Handlingar, Band 22, no. 8, p. 5.
[7] Skey. Mineral Waters of New Zealand · Trans. New Zealand Inst., vol. 10, p. 433.

At the Azores, Mr. Mosely, naturalist on the *Challenger* expedition, found algæ growing about the hot springs of Furnas Lake, island of St. Michael.[1] The algæ occur on areas splashed by the hot sulphurous waters, forming a pale, yellowish-green layer an inch and a half thick. The color is most intense in the inner layers of the growth. This gelatinous vegetable matter occurs mingled with a gray earthy material in successive layers. The temperature of the water was 176° F. to 194° F. A thick brilliant green deposit, consisting of *Chroococcus*, was found at the edge of a shallow pool of hot water whose temperature was between 149° F. and 156° F. Specimens were also collected from a swamp of hot mud in which, beside algæ, a rush (*Juncus*) was found growing. The temperatures given by Mr. Mosely are all estimated, but are probably correct within the limits stated. The specimens obtained from these springs were examined by Mr. W. Archer, and the result of his study published in a paper[2] which is mainly botanical, but is interesting in this connection as showing the identity of many of the species from the hot waters of the Azores, with species common in the cold waters of Great Britain.

In the narrative of the voyage of the *Challenger*, Mr. Mosely describes the occurrence of similar hot-water growths at the Banda Islands and at the new volcano of Camiguin. At the first locality gelatinous masses of algæ resembling those growing in the Azores hot springs were found around the mouths of fissures from which jets of steam issued, the only water present being that supplied by the condensation of the vapor. This sulphurous steam had a temperature of 250° F. within the fissure, and the thermometer stood at 140° where the algæ flourished. In some places the algæ and a white mineral incrustation formed alternating layers.[3]

At the base of the new volcano of Camiguin two hot streams were full of algæ. No vegetation was found in hot water where the temperature exceeded 145°.2 F., but in the stream-bed green patches occurred on stones projecting above the surface. As the water of this stream became cooler, a few yards farther down, algæ were found growing in the middle of the channel at 113°.5 F. "This," says Mr. Mosely, "seems to be the limiting temperature for this particular algæ in this water." Where the temperature of the stream was 122° F. it fed a little side pool where algæ were growing at a temperature of 101°.5 F.[4]

Dr. Hooker found algæ in the hot springs of the Himalayas at several localities.[5] At Soorujkoond or Belcuppec (in the Behar Hills)

[1] Journ. Linn. Soc. (Bot.), vol. 14, 1875, p. 321.
[2] Ibid., p. 328.
[3] Voyage of H. M. S. *Challenger*: Narr., vol. 1, part I, p. 563.
[4] Ibid, p. 654.

brown and green algæ were found forming broad, luxuriant strata
where the temperature did not exceed 168° F., the growth thriving
until the water had cooled down to 90° F. The brown algæ were
found in deeper and hotter water than the green. In an appendix,
Rev. J. M. Berkley, writing about the specimens collected from these
springs, says a *Leptothrix* occurs from 80° F. to 158° F., an imperfect
Zygnema between 84° F. and 112° F.² The same writer also describes
specimens collected by Dr. Hooker from the hot springs of Momay
at 110° F., and from Pugha, Thibet, in springs having a temperature
of 174° F.³

Green algous growths were observed by Prof. J. D. Dana in the
hot springs of Luzon at 160° F.,⁴ and similar vegetation was found
in the Celebes hot springs in waters of 123°.8 F. and 170° F. by Prof.
A. S. Bickmore.⁵ Green algous vegetation was also noted in many
Japanese hot springs by Benj. Smith Lyman.⁶ Cushion-shaped
masses of slippery gelatinous vegetation—*Oscillatoria labyrinthi-
formis* Ach.—were found by Junghuhn in the hot springs of Java⁷ at
150° F., and a species of *Oscillaria* was found associated with a milk
white precipitate of *sulphate* of lime at Tji-Panas.

In the United States algæ have been found in hot springs at many
localities. Red, white, and green growths occurring in the warm
sulphur springs of Virginia have given rise to the names of many of
these famous resorts.⁸ At the hot springs of Arkansas green cryptogamic vegetation occurs in water having a temperature of 140° F.,
and a species from this place is described by Kützing.⁹

At the so-called "geysers" of Pluton Creek, California, green algæ
occur in the hot acid water in great abundance. Prof. W. H.
Brewer found that the highest temperature noted at which the plants
were growing was 200° F., but they were most abundant in waters
of 125° F. to 140° F. The growth in the hottest waters was apparently of the simplest kind, and composed of simple bright-green cells.
Where the water had cooled to 140° F. to 149° F., bright-green filamentous confervæ formed in considerable masses. Similar growths
formed coatings on the soil about steam-jets, and were alternately
exposed to very hot steam and cooler air. In a specimen collected
from water having a temperature of 130° F., Mr. A. M. Edwards, of

¹ Himalayan Travels: Jos. Dalton Hooker, vol. 1, pp. 27, 379.
² The temperatures given by Mr. Berkley are 10° lower than those given by
Hooker.
³ Loc. cit., p. 379.
⁴ Manual of Geology, by Jas. D. Dana: 3d ed., 1880, p. 612.
⁵ Travels in East Indian Archipelago.
⁶ Prelim. Reports, Geol. Surv., Japan, 1874, 1877, 1879.
⁷ Java, Seine Gestalt, Fr. Junghuhn: vol. 2, pp. 864, 866, 868, 870, 873.
⁸ Geology of the Virginias, W. B. Rogers, pp. 107, 589.
⁹ Species Algarum.

New York, is said to have found animal as well as vegetable organisms. Professor Brewer also states that in the hot siliceous waters of Steam-boat Springs, Nevada, there is an abundant confervoid vegetation in the gelatinous mass formed where the water spreads out over the surface. This was most plentiful where the temperature was about 100° F. The most interesting feature of this occurrence is the abundant vegetation in the gelatinous silica.[1] Mr. James Blake found diatoms in water having a temperature of 163° F. at the Pueblo Hot Springs, Nevada.[2] The algæ growing in the Benton Spring, Owen's Valley, California, are described by Mrs. Partz as representing three forms. The first is developed in the basin of the spring at a temperature of 124° F. to 135° F., the temperature of the water at the point of issue being 160° F. In the warm creek formed by the overflow of the spring the algæ form waving fibers two feet long, at temperatures between 110° F. and 120° F. Below 100° F. these plants cease to grow, but a bright-green, slimy fungus occurs, disappearing as the temperature decreases. Dr. H. C. Wood gives technical descriptions of these plants,[3] and says the forms develope at the highest temperatures are immature. The presence of green confervoid vegetation in many other hot springs has been noted by various writers, but no description has been given either of the plants themselves or of the temperature and other conditions governing their occurrence.

In the hot springs of the Yellowstone National Park no plant life has been found at a temperature exceeding 185° F., but at temperatures between 90° F. and 185° F. algous growths are generally present. In the reports of the Hayden Survey for 1871 and 1872 there are several references to the presence of vegetation in the hot waters. At the Mammoth Hot Springs, Dr. F. V. Hayden observed the occurrence of pale yellow filaments about the springs and the green confervoid vegetation of the waters, as well as the presence of diatoms in the basins of the main springs, two species of the latter, *Palmella* and *Oscillaria*, being recognized by D. Billings.[4] Green vegetation was also noted in the hot waters of the Washburne, Pelican, and Terrace Springs, and at the Lower Geyser Basin.[5] The brown leathery lining of the springs of the Lower Geyser Basin of the Firehole River was thought by Dr. Hayden to be an aggregation of diatoms covered with iron oxide.[6] In 1872 Prof. F. H. Bradley recognized the presence of vegetation in the hot springs of the park, and writing of the hot waters of the Geyser Basins, says

[1] Amer. Jour. Sci., 2d series, vol 41, p. 391.
[2] Manual of Geology, by James D. Dana, 3d ed., 1880, p. 611.
[3] Amer. Jour. Sci., 2d series. vol. 46, p. 31.
[4] Ann. Rept. U. S. Geol. and Geogr. Survey of the Territories for 1871, pp. 69,70.
[5] Loc. cit., p. 136, and 1872 report, p. 55.
[6] Loc. cit., 1871, p. 105.

that there are gelatinous forms allied to mycelium, or mother of
vinegar, in nearly all the pools, except where the ebullition is so
strong as to break up such tender tissues. This material occurred in
broad, thick sheets of green or rusty brown, in thick, branching
forms, resembling sponges, or in long waving white fibers.[1] In the
mucilaginous deposit on the side of a spring at the Lower Geyser
Basin Dr. Josiah Curtis found skeletons of diatoms, but no living
ones. Professor Bradley said the colors striping the mound of the
Solitary (Lone Star) Geyser are due to purely vegetable material.
His assistant, Mr. Taggart, reported leafy vegetation in springs of
120° or less at Lewis Lake, where the springs of higher temperature
contained pulpy masses of a fungoid growth common about the hot
springs of the Geyser Basins.[2] The botanist of the expedition,
Prof. John Coulter, says in his report that algæ were discovered
growing in some of the hot springs. He collected orange-colored
confervoid specimens from the waters of the Lower Geyser Basin
which were identified by Charles H. Peck as *Confervæ aurantica.*[3]
Prof. Theodore Comstock, who visited the park with the Jones expedition in 1873, records the presence of green confervæ in the Green
Spring, Pelican Creek, at 104° F., and similar growths were found
at Turbid Lake, Mammoth Hot Springs. Excelsior Geyser, Sapphire
Springs, and the Lake Shore Hot Springs.[4] The same observer
noticed the silken yellow filaments at the Mammoth Hot Springs,
and supposed the abundant "colloid matter" of the springs to
originate from organic matter contained in the water, the forms
being produced by the rising or buoyancy of bubbles of carbonic
acid gas.[5] Dr. C. C. Parry, botanist of this expedition, noticed the
presence of algæ in the hot springs of the park, and says they will
reward special research.[6]

The report of Dr. A. C. Peale upon the thermal springs of the
Yellowstone National Park[7] contains but little about the vegetation
of the Park hot springs. Under the heading, " Life in Hot Springs,"
he says :

At numerous places in all the geyser areas and at Gardiner's River masses of
gelatinous material of greenish-red, yellow, and brown colors are noticed, and usually have been considered of organic origin. In most cases where microscopical
examination has been made no trace of vegetable organization has been noted, and
in regions where the springs are siliceous this curious material is probably that form
of gelatinous silica described in another place as viandite. In some springs of very
low temperature a brown, leathery-looking material is found lining the basins. It

[1]Ann. Rept. U. S. Geol. and Geog. Survey of the Territories for 1872, pp. 207, 231.
[2]Loc. cit., 1872, p. 250.
[3]Loc. cit., p. 752.
[4]Report of Reconn. N.W. Wyoming in 1873, by Capt. Wm. Jones, U. S. War
Dept., pp. 190, 194, 210, 228, 231, 238.
[5]Loc. cit., p. 207.
[6]Amer. Naturalist, 1874, p. 178.
[7]Final Rept. U. S. Geol. and Geog. Survey Terr., 1878, vol. 2, p. 359.

becomes hard upon drying, but has not as yet been examined microscopically or chemically, so that its nature is unknown, but in all probability it is one of the forms of silica rather than an organic material.

This evidently represents the views of Dr. Peale and his colleagues regarding the nature of the algous growths of the Park hot springs at the time this report was prepared.

This review of the literature of the subject shows how few occurrences of hot-spring vegetation have as yet been carefully observed and described. In the cases noted, naturalists have generally given the temperature of the water in which the plants were found, and the specimens collected have been studied from a purely botanical point of view, but with the notable exception of Prof. Ferd. Cohn, observers have entirely overlooked the geological work of the lowly organized plants and the part they take in the production of hot-spring deposits.

These hot-water growths, like all fresh-water algæ, are more widely distributed than any other plants save those peculiar to brackish waters. This is shown by the occurrence of algous vegetation in the hot springs of such widely separated localities as Iceland, the Azores, New Zealand, Japan, and the United States. A comparison of the species shows that the flora is very uniform in character, being limited to a few groups and the species themselves being identical to a great extent.

Perhaps the most interesting feature connected with the life of these algæ is their tolerance of a high degree of heat. The extreme temperature at which vegetation has been observed is 200° F., recorded by Prof. W. H. Brewer at the California "Geysers." On the island of Ischia, near Naples, no algæ were found in hot waters above 185° F., which accords with the observations made in the Yellowstone National Park. At other places these growths have not been found at such high temperatures. Dr. J. D. Hooker found the limit to be 168° F. in the Himalayan springs, and Prof. Ferd. Cohn says no growths are present in the Carlsbad waters, where the temperature exceeds 44° R. (131° F.). As regards the effect of the chemical substances dissolved in the water there is but little known, but vegetation has been found in all varieties of water, sulphurous, calcareous, acid, and alkaline, and so far as observed the amount of material held in solution does not affect the growth. Certain species, however, are known to be peculiar to particular waters. Thus the *Beggiatoa* form the characteristic vegetation of sulphur springs, and *Gaillionæ* are found in iron-bearing waters. The adaptability of particular algæ to extreme conditions of environment is shown by the occurrence of the same species in the highly heated sulphurous-siliceous waters of the Azores and the cold surface waters of Great Britain.

Altitude is not known to affect the growths, and algæ are found in Iceland but a few hundred feet above sea level, in the Yellowstone National Park at 7,500 feet, and in the Himalayas at an elevation of 17,000 feet.

HOT SPRINGS OF THE YELLOWSTONE NATIONAL PARK.

Regions of solfataric activity have always been of peculiar interest to scientific observers, not only on account of the curious and often extremely beautiful hot springs and the rarer occurrence of geysers in such districts, but also from the varied phenomena of rock decomposition and of mineral formation and deposition which always accompany such hydrothermal action.   It is in these natural laboratories that we are permitted to see in operation processes which have produced important changes in the rocks of the earth's crust and afford a key to many of the problems of chemical geology.

There is perhaps no other district in the world where hydrothermal action is as prominent or as extensive as it is in the Yellowstone National Park.   In this area of about 3,500 square miles, over 3,600 hot springs and 100 geysers have been visited and their features noted, and there are also almost innumerable steam vents.   With few exceptions the hot waters are siliceous, and rise through the acidic lavas of the park, and it is probable that it is owing to this fact that the deposits formed by the hot waters do not differ more in character.   The facts upon which this paper is based have been obtained in the course of a series of comparative observations carried on by the writer for the past six years at the different hot-spring areas of the Park, under the direction of Mr. Arnold Hague, geologist in charge of the Geological Survey of the Yellowstone National Park.

THE MAMMOTH HOT SPRINGS.

Although the Yellowstone Park abounds in hot springs, calcareous hot waters are extremely rare, and but one locality is known where such springs have formed deposits of travertine, or calcareous tufa, of any considerable extent.   This is the Mammoth Hot Springs.   At this place the heated waters rising through Mesozoic limestone reach the surface heavily charged with carbonate of lime in solution, which is deposited by the hot waters in the form of travertine, affording an excellent opportunity for a study of the formation of this mineral.

Calcareous hot waters are not rare in nature, but are found in many parts of the world, and are usually surrounded by deposits of travertine often of considerable extent; yet there are few places where such deposits equal those of the Mammoth Hot Springs in magnitude, and none exceeding them in beauty.   The travertine deposits of Hierapolis in Asia Minor, famous for its hot waters in the time of the Emperor Constantine, form a white hill whose slopes are ornamented with basins resembling those of the Marble Terrace of the Mammoth Hot Springs, and the springs of the Hammon Meschoutin,

TERRACED BASINS OF BLUE SPRINGS, MAMMOTH HOT SPRINGS.

in Algeria, have built up cones and ridges which are the duplicate
of those found on the terraces of our own locality.

GEOLOGICAL RELATIONS.

The Mammoth Hot Springs form the most northern of the numerous hot-spring areas of the Park, being situated in the northwest
corner of the reservation, three-quarters of a mile south of the forty-
fifth parallel, which forms the Montana-Wyoming boundary. As it
is but seven miles from the terminus of the railroad it forms the first
stopping place of the traveler who enters the Park from the north,
and it is the most accessible of the many points of interest in this
region. The situation is extremely picturesque; the dark and lofty
summit of Sepulchre Mountain rising near by on the north, while the
upper valley of the Yellowstone and the sharp peaks of the Snowy
Range are seen at the northeast, between the slopes of Sepulchre
and the long mural face of Mount Evarts. In the southeast the eye
dwells pleasingly upon the distant view of the ravine of Lava Creek
and Undine Falls, with many snow-flecked peaks in the far distance.
Bunsen Peak rises abruptly in the south, its dark slopes forming a
pleasing background to the white mass of hot-spring deposit when
seen from the north. This deposit fills an ancient ravine lying between Terrace Mountain and Sentinel Butte, the grassy slopes of the
latter showing exposures of Jurassic and Cretaceous limestones
carved into well-defined benches by glacial action. Immediately
south of the travertine terraces the sedimentary strata are covered
by rhyolite, the northern extension of the great lava flows which fill
the ancient basin of the Park. Near the Gardiner River, Cretaceous
sandstones form small ridges, dividing the travertine sheet into
three tongues; these beds dip steeply eastward, passing beneath the
strata forming the face of Mount Evarts.

TRAVERTINE DEPOSITS.

The total area covered by the travertine is about two square miles,
including the beds of preglacial age which form the summit of Terrace Mountain. The greatest thickness is probably about two hundred and fifty feet, but the average is very much less. The upper
limit of the deposit, forming the terraces and filling the ravine, is
about 1,400 feet above the Gardiner River and 7,100 feet above sea
level; the travertine extends from this terrace down to the river,
forming a continuous covering of varying width and thickness. It
is impossible to measure the volume of the deposit as the thickness
is variable, and the contour of the underlying surface can be con-
jectured only by the relation of the neighboring slopes.

The usual approach to the Mammoth Hot Springs from the railroad is over the road leading up the picturesque gorge of the Gardiner to the foot of the terraces. Recrossing this stream near its

junction with the Hot River, the road gradually ascends to the flat or terrace on which the hotels stand, 500 feet above the river. The road is built upon the hot-spring deposit, hidden on the lower slopes by drift and soil but exposed during the last 200 feet of the ascent, where many well-preserved basins may be seen on the pine and cedar covered slopes.

When first seen the main mass of the recent deposit is striking from its whiteness, resembling an immense snow-bank, filling a narrow valley whose pine-clad sides are in strong contrast to the white travertine. It has been compared by Prof. Arch. Geikie to the terminal front of a glacier, and by other writers to a foaming cascade suddenly turned into stone. Streaks and patches of red, yellow, and green seen upon these white slopes mark the course of the overflowing waters, and clouds of steam float lightly upward from the springs of the main terrace and vanish in mid-air. There are in all eight well-defined benches or terraces formed by the travertine, each with a more or less level surface, and terminated by steep slopes leading to the terrace below. The largest of these flats is the Hotel terrace, which is 83 acres in extent. This possesses several features of interest. These are usually overlooked in the desire to see the greater wonders and beauties of the upper terraces, but one can scarcely fail to notice the Liberty Cap, a pillar 43 feet in height with sphinx-like profile, the cone of a hot spring long extinct. This cone and its companion, the Thumb, with the immense empty hotspring bowls of this terrace, attest an activity and size for these extinct springs far surpassing any now active.

THE SPRINGS AND THEIR VEGETATION.

With the exception of the Hot River all the active springs now issue from the terraces above the hotels, or from the upper part of the hotel terrace itself. These seventy-five springs vary in temperature from 80° F. up to 165° F., and in size from small oozes of hot water to basins 50 by 100 feet across, with an overflow of many thousand gallons per hour. Algæ have been found in all these springs, and it is this vegetation, and the part which it takes in the formation of travertine by the hot waters, that are of especial interest in the present paper.

In wandering around the terraces of this great deposit of travertine the observer is sure to be impressed with the brightly tinted basins about the springs and the red and orange colors of the slopes overflowed by the hot waters. These colors are due to the presence of microscopic algæ, which are not easily recognizable in this deposit, owing to their covering of travertine. In the cooler springs and channels similar vegetation forms the bright green, orange, or brown membrane-like sheets or masses of jelly, without apparent vegetable structure.

The true nature of the silken yellow filaments found in the bowls and channels of even the hottest springs is more apparent, though the yellow color is due to sulphur incrusting the algæ threads. The intimate relation of these algous growths to the deposits of newly formed travertine suggests at once that the algæ are encrusted by the carbonate of lime, and so aid in the formation of the tufa. While this is probably true, the chief work of these plants is the separation from the water of the carbonate of lime, which they cause by their abstraction of carbonic acid. Owing to this action, a common function of vegetation, such growths are an important factor in the formation of travertine by the Mammoth Hot Springs waters.

GENERAL OCCURRENCE OF THE ALGÆ.

The general occurrence of the algous vegetation will be best understood if a brief description of a few of the typical springs is given.

The largest springs now active are those of the Main Terrace. This is a fairly flat area of 8¾ acres in extent and 250 feet above the hotels. On the north the terrace ends in abrupt slopes, extending down to the bench below ; on the east and southeast the descent is more gradual, extending down to the military quarters 175 feet below Near the center of this terrace are the Blue Springs. These springs shift their position from year to year, the rapid deposition of travertine choking up the vents, causing the springs to seek other and easier outlets. In this case it often happens that the pressure of the accumulated gas fractures the deposit, and the water issues in a jet a foot or more in height. A rim of travertine is soon built up about the vent, forming a basin, into which the water, now relieved of the excess of pressure, issues quietly, though in considerable volume. The most beautiful of the Blue Springs is a pool 15 by 20 feet in extreme dimensions filled with pellucid water apparently in violent ebullition. The sides and bottom of the basin are formed of pure white travertine, while the varying depths cause the water to appear all shades of blue and green, from a deep peacock blue in the deeper parts of the bowl to the lightest of Nile greens in the shallower recesses. The water, issuing with a temperature of 165° F., contains a large amount of gas, which escapes at the surface of the pool, causing the water to rise in a low dome, variations in the amount of gas producing a pulsating movement, sending out waves which ripple across the water and curl over the shallow margin of the bowl. The overflow passes over and under large fan-shaped masses of fibrous white or yellow travertine (Fig. 52) into the uppermost of a series of basins irregularly arranged in tiers, a portion running in serpentine waterways built up of travertine. These natural aqueducts are often two or three feet high. In the center is a

632   FORMATION OF HOT SPRING DEPOSITS.

shallow gutter too small to hold the volume of the stream, which overflows the sides and fills the basins along its course.

Fig. 52. Travertine fan, main terrace, Mammoth Hot Springs.

These terraced overflow basins form the most striking feature of the springs. No description can do justice to their beauty, for neither the delicate fretwork of their walls nor the frosted surface of the glistening deposit, nor the brilliant colors of the pools and rims can be described. Plate LXXVIII, from a photograph, shows a few of the many basins, of which each differs from the others. The walls are built up of pure white travertine, the surface resembling imbricated shells or a multitude of miniature basins, and often covered with a brightly colored vegetable jelly where the water is slightly cooled. These basin walls vary in height from a few inches to several feet. Their outline is rarely crescentic, usually irregular, wavy, and scalloped. The water runs over the rims in thin sheets and little cascades, depositing travertine wherever it flows and constantly building up the basins until the flow is checked by the increased height of the rims. Yellow sulphur-coated algæ threads are abundant in the bowl of the spring and the rapid-flowing streams, but the exquisite blues and greens of the hottest basins are due solely to the varying depths of water. The bright lemon, red, and green shades of the cooler pools are, however, entirely vegetable in their nature, and due to the presence of algæ lining the basins and striping their outer walls. In a general view of the entire collection of these basins, obtained from the edge of the terrace above, the effect is that of a brilliant mosaic, the colors occurring in well-defined areas outlined by the white travertine rims. As will be shown later, the contrasting tints of adjacent basins are due to the different temperature of the water and consequent different development of the algous vegetation. Looking at the pools near by proves that these colors are not pure, but are produced by a number of tints, minute differences in depth producing variations in color in the same basin. Large as is the overflow from the Blue Springs, little reaches the edge of the terrace, the water sinking into the porous deposit or flowing into holes and fissures in the travertine floor.

On the same terrace, but close to the southeastern edge, are the

MARBLE BASINS, MAMMOTH HOT SPRINGS.

two main springs. They are very much alike, and are to-day in
nearly the same condition as in 1871, when they were first seen.
The northern spring is a brown lined bowl, 75 by 100 feet across, and
5 to 8 feet deep. The flat margin is formed of smooth and polished
salmon-colored travertine whose thin laminæ and hardness show it
to have been quite slowly formed. The water is much cooler than
that of the Blue Springs, having a temperature of 136° F. at the edge
of the bowl. The supply is constant and issues from holes in the
bottom of the basin, their location being distinguished by the lighter
color of the water, the eddying currents, and an occasional stream
of gas bubbles.

The perfect transparency of the water enables one to see the mi-
nutest details of the sides and bottom of the bowl. The volume of
water which the two main springs pour out is not known, as the outflow does not run in definite channels, but pours over the eastern
margins in a shallow sheet which, spreading out, flows down the
rippled slopes and over the Marble Basins. Pl. LXXIX shows a few
of the upper basins, which are often quite shallow, and hardly merit
the name of basin. Here the waters deposit carbonate of lime rapidly, and the walls or basin-fronts are generally solid, while on the
lower slope the cooled waters have parted with much of their lime,
and deposit travertine slowly. On these lower slopes the basins are
fringed with slender stalactites and pillars, forming the beautiful
Pulpit Basins, illustrated in Pl. LXXX.[1] In this case, also, the pool
or basin proper is very shallow, rarely a foot deep, and the rim or
lip generally projects over the pillared front, as it is here that the
deposition of travertine is most rapid.

Wherever the hot waters flow the deposit is brightly colored by
the algous vegetation. The pools and basins near the springs are
lined with deep red, while the slopes below are bright orange in
color, and it is only near the base of the slope, where the waters are
quite cool, that this color disappears.

EFFECT OF ENVIRONMENT.

Algæ are abundant in all the springs and wherever the hot waters
flow, but the growths vary in character and in color with the temperature of the water and with the situation in which they occur.
If the temperature exceeds 150° F. a white filamentary alga is the
only species present, the thread generally coated with sulphur; but
where the water has cooled below that point, or issues with a lower
temperature, a pale greenish-yellow growth is found, sparingly at
first, but more abundantly and deeply tinted in cooler water, where
it often entirely replaces the white species. This green alga is associated in turn with a red or orange species which gradually replaces

---

[1] Plates LXXVIII, LXXIX, and LXXX are engraved from photographs made by
T. W. Ingersoll, of St. Paul, Minn.

it in the cooling waters, while in tepid streams too cool to support any of these forms an olive-brown species forms a soft, velvety covering over the travertine. Different conditions of flow and current produce varying forms of the same growth. In a rapid current the algæ are filamentary, while in quieter water the threads are united together in a membrane-like sheet or in masses of jelly, generally inflated by gas bubbles entangled in the vegetable tissue. At the borders of many channels the two forms pass into each other. Where the deposition of travertine is very rapid, as is generally the case on the overflow slopes and basin walls, the algæ are encased in the deposit and only the vegetating ends of the filaments are exposed and free.

The white algæ, which grow in the hotter waters, are generally coated with sulphur near the source of the spring, forming tufts of bright yellow filaments resembling skeins of silk, vibrating with the eddies and currents of the stream. Farther from the source these threads are not sulphur-coated, but are encrusted with carbonate of lime, and they form the radiating, fan-like masses of travertine shown in Fig. 52. The white algæ are generally found in the

FIG. 52. Coating specimens, Mammoth Hot Springs.

rapid currents of overflow streams; rarely in the eddying waters of the hot-spring bowls.

It was suspected that this white or sulphur-coated species, so abundant in the hotter waters, might be identical with other and more brightly colored algæ, which had not been bleached by the hot sulphurous water. Specimens of a dark emerald-green growth were therefore placed in the overflow of a hot sulphur spring alongside of the white sulphur-coated filaments. In a few hours the green color had disappeared from the submerged portion of the green growth, and the white bleached filaments were partially coated with sulphur. Subsequent observations proved, however, that the white species maintains its character in comparatively cool waters, where it occurs associated with red and with green algæ, so that the experiment does not show the identity of the species, as was at first supposed. The green algæ are not such active travertine formers as the white and red species. They thrive best in the shade, or where hidden from the light by a covering of the red algæ, and the rich emerald-green color of the species changes to an olive or dull brown where exposed to direct sunlight. Flowing water seems to be necessary for its best development, so that, unlike the red algæ, the green species is seldom found on the bottom of overflow pools and basins. In water too hot for the full development of this alga, the growth is pale, yellowish green in color, or even bright yellow, frequently occurring in gelatinous masses showing no trace of filaments. In cooler water the color deepens to a rich emerald.

The orange or red algæ are very active in the formation of travertine, and there is not an overflow slope that does not show traces of its color. It tints the bottoms of the basins about the springs, and their rims and walls, with its varying shades of yellow, red, or brown, and it is this growth that imparts the tawny yellow or orange color to the slopes about all the springs, particularly noticeable on the slopes overflowed by the waters of the Main Spring. This species is rarely found free from lime, which generally incrusts it so thickly that it is difficult to distinguish its vegetable nature.

DESCRIPTION OF THE VEGETABLE GROWTH.

The springs at the foot of the slope below the Pulpit Basins formerly presented the most luxuriant algous vegetation of the locality, in the area overflowed by their waters. This overflow is now chiefly used to supply the military stables, but formerly covered the flat north of the springs, flowing over a cushion of algous jelly several inches thick. The temperature of the spring was 127° F., and the channel near it was filled with a bubbly, gelatinous vegetation, emerald save at the edges, where it shaded into a dull, greenish brown. This changed gradually to a mixture of green and yellow where the temperature had fallen to 115°, but at 110° the surface of the jelly showed no trace of green, and the orange species only was seen, continuing abundant until at 97° F. the growth ceased. Travertine deposition was taking place most rapidly where the temperature was 100° F., and the algæ were so heavily coated and inclosed by the deposit that its organic character was completely hidden, the light salmon-red color being apparently due to some mineral. It should be mentioned in this connection that these springs contain much less lime in solution than the other springs of this locality, and the waters do not coat and incrust articles placed in their spray. Only a small part of the overflow seems to have run over this salmon-colored crust, for upon tearing off this coating the inside is seen to consist of green algæ, the larger part of the overflow running through this vegetable conduit.

The association of the different species is well illustrated in their occurrence at the Orange Spring. This vent has built up a mound 15 feet high, 20 to 25 feet wide, and 50 feet long, with a gently arched summit and steeply sloping sides. The water issues from several little cones on the summit, situated along a line corresponding to the major axis of the mound, but there is usually one vent from which the greatest part of the overflow issues, and generally in a jet nearly a foot high. The temperature is but 148° F., so that this spouting is due to gaseous or to hydrostatic pressure. Falling into a little basin the water flows off in a ramifying network of little channels to the edge, where spreading out it forms a thin, glistening sheet, dashing and rippling down the steep slopes, only to sink in the porous travertine at the base of the mound. The water has a strong sulphurous odor, and the deposit about the vent contains considerable sulphur. If under water the surface of this deposit is black, with bunches of sulphur-coated filaments attached to the sides and bottom of the channel. Near the vent these are the only algæ present, but pale yellowish green threads are found in the cooler water at the border of the channel, and are abundant at 130° in many of the branch streamlets. As the water cools still more the green growth becomes deeper in color, and the red species appears at the edge of the stream. This growth is sometimes filamentary, but generally a jelly-like membrane, when not buried in the travertine. The surface of the mound between the reticulated channels is covered with a gelatinous coating of red and green algæ similar to those just mentioned, but mixed with crusts of carbonate of lime. Mushroom-shaped forms of salmon-colored travertine rise from the bed of the larger channels or project over the edges of the stream; these are formed partly by algæ and partly by evaporation.

The steep rounded or step-like slopes of the mound are bright orange in color where covered by the water, and it is undoubtedly this which gave the name to the spring. This coloring is due to algæ similar to those found on the summit of the mound, but the filaments are buried in the travertine, their tips alone projecting, reminding one of the growing points of peat mosses whose stems can

PULPIT BASINS, MAMMOTH HOT SPRINGS.

be followed down into the peat beneath. Where the overflow has
become too cool to support the orange algæ, which does not live
below 90° F., a brown species forms a smooth velvety coating on the
travertine, and is very abundant at 85° F.; this in turn disappears as
the water becomes still cooler.

In the overflow basins of the Blue Springs (Plate LXXVIII) the
colors of adjoining pools are often quite different. In one of the hotter basins where the temperature was 142° the algæ tinting the deposit
were a bright lemon yellow, while the rich, deep red growth of the
adjoining basin lived at 115° F. The red growth is very prominent
in water between 110° F. and 130° F. At the edge of a pool where the
flow was comparatively quiet, a pistachio green growth merging into
yellow and orange began at 145° F., the growth being thin and closely
adherent. Close by a place where the flow was very sluggish, at a
temperature of 130° F., the orange algæ are abundantly developed
in gelatinous balloon-like forms. At 115° F. the red tint is much
browner and at 95° F. is a dark orange brown.

In several of the basins, yellow, red, salmon, green, and brown
interblend, owing to differences in depth and consequently in temperature and current. The vegetable nature of such growths is generally much obscured by the accompanying deposition of travertine.

SOLUBILITY OF CARBONATE OF LIME IN NATURAL WATERS.

The large amount of carbonate of lime which the hot waters of the
Mammoth Hot Springs contain in solution suggests an inquiry regarding the conditions under which such waters take that salt into solution.

In pure water the carbonate of lime is very sparingly soluble, the
proportion given by Bineau being but one part in 30,000, to one in
50,000, or according to Fresenius, one part in 10,800 cold and 8,875
parts of boiling water. In carbonated waters the neutral carbonate
of lime unites with the carbonic acid to form the bicarbonate of
lime, which is readily dissolved in water to the extent of 0.88 grammes
per litre in water saturated with carbonic acid gas at the ordinary
atmospheric pressure and a temperature of 10° C. With an increase
of pressure the amount taken into solution is augmented with the
increase of carbonic acid absorbed, but the maximum amount that can
be dissolved is about 3 grammes per litre.[1] The presence of alkaline
and earthy salts in water free from carbonic acid favors the formation of unstable supersaturated solutions of carbonate of lime, from
which the lime is gradually precipitated, this separation being more
rapid from waters containing the chlorides than from those holding
the sulphates of the alkalies and the alkaline earths. Magnesium
sulphate and sodium sulphate form solutions with a certain amount
of stability, but the lime is all precipitated in eight to ten days.[2]

[1] Roscoe and Schorlemmer, vol. 2, p. 308.
[2] T. Sterry Hunt : Am. Jour. Sci., 2d series, vol. 42. p. 58.

In water saturated with carbonic acid the alkaline and earthy chlorides form unstable supersaturated solutions, from which the lime soon crystallizes out as the hydrous carbonate (at low temperatures) and the solution then contains but 0.8 grammes of carbonate of lime per litre, corresponding to that dissolved by the carbonic acid.[1] But the capacity of carbonated waters for carbonate of lime is nearly doubled by the presence of magnesium sulphate or sodium sulphate in the solution. Water holding either of these sulphates in solution in the proportion of $\frac{1}{1000}$th or even less, and impregnated with carbonic acid, readily takes into permanent solution at the ordinary temperatures and pressure a quantity of pure carbonate of lime equal to 1.56 to 1.82 and even 2 grammes to the litre.[2] It is thus evident that solutions of carbonate of lime in pure or mineral waters are permanent only in the presence of free carbonic acid.

CHARACTER OF THE HOT SPRING WATERS.

The water of the Mammoth Hot Springs is remarkably clear and transparent; the temperature varies at different springs from 80° F up to 165° F., exceeding 130° in all the larger springs. While hot it generally possesses a sulphurous odor, the intensity varying greatly at different springs, but always being strong if the temperature exceeds 140°, when sulphur is found incrusting the algæ filaments growing near the vent of the spring. When cold the water is not peculiar in taste or in odor, but it is considered unfit for drinking, owing to the large amount of carbonate of lime which it holds in solution.

At many of the springs a large amount of gas escapes as the water issues from the vent, which is proven by analysis to consist of carbonic acid gas, oxygen, and nitrogen. Although the odor of sulphur is very noticeable and sulphur is deposited at many of the springs, the amount of sulphuretted hydrogen present in the water is very small, and is too minute to appear in the analysis of the waters. The general character of the water is the same in all the springs, but the amount of mineral matter held in solution varies at different springs from 15 to 17 parts in 10,000, and of this one-third consists of carbonate of lime and the remainder of readily soluble salts.

In the following table analyses are given of typical waters from the Mammoth Hot Springs, and also of the surface waters of the surrounding slopes. These analyses were made by Prof. F. A. Gooch and J E. Whitfield,[3] for the Geological Survey of the Yellowstone National Park. In the same table analyses for comparison are given of the thermal waters of Hierapolis and Kukurtlu,

[1] Hunt. loc. cit. and Skey, in Trans. New Zealand Inst., vol. 9, p. 454
[2] Hunt, loc. cit., p. 50.
[3] Analyses of Waters of the Yell. Nat. Park, Bull. No. 47, U. S. Geol. Survey.

Asia Minor, made by J. Lawrence Smith,[1] and also of the **Carlsbad** sprudel, made by Ragksy.[2]

Analyses of waters from the Mammoth Hot Springs.

	Cleopatra.	Orange.	136° F. Hot River.	Gardiner.	Hotel water.	130° F. Hierapolis.	182° F.	Carlsbad.
Ca Cl₂	0.0009	....	....	....	....	0.090	....	....
NH₄Cl	0.0019	....	0.0003	....	....	....	....	....
Li Cl	0.0140	0.0007	0.0008	Trace.	....	....	....	....
Na Cl	0.1903	0.1696	0.1855	....	....	....	....	1.0306
K Cl	0.0976	0.1165	0.0882	0.0103	0.0046	....	....	....
K Br...............	Trace.	Trace.	....	....	....	....	....	....
Na₂ SO₄	0.1448	0.1834	0.2205	0.0161	0.0448	0.341	0.1956	2.3721
K₂ SO₄	....	....	....	0.0036	0.0015	....	0.0202	0.1636
Mg SO₄	0.3645	0.3295	0.3155	....	0.0076	0.431	....	....
Ca SO₄	0.1933	0.2902	0.1450	....	....	0.119	0.1710	....
Al₂ (SO₄)₃	....	....	....	....	....	....	0.0043	....
Na₂ B₄ O₇	0.0026	....	0.0185	....	....	....	....	....
Na As O₃	0.0041	....	0.0004	....	....	....	....	....
Ca CO₃	0.6254	0.5580	0.4833	0.0623	0.0790	*1.308	*0.1832	0.2978
Mg CO₃	....	....	....	0.0018	0.0258	*0.041	*0.0460	0.1240
NaH CO₃	....	....	....	†0.6340	....	0.078	....	†1.3619
Fe CO₃	....	....	....	....	....	....	....	0.0028
Mn CO₃	....	....	....	....	....	....	....	0.0006
Al₄ O₃	0.0063	.0022	0.0007	0.0079	0.0021	....	....	‡0.0004
Si O₂	0.0347	0.502	0.0600	0.0469	0.0355	0.008	0.1100	0.0728
Total solids...	1.7315	1.6183	1.5297	....	....	1.934	0.970	5.4312
Total CO₂ §	0.3537	.0924	0.2143	0.0296	0.0748	(0.5320)	....	0.7694
Summation....	2.0832	1.7057	1.7449	0.3137	0.2757	....	....	....

* Bicarbonate.   † Neutral.   ‡ Al₄ (PO₄)₃.   § Free.

The analyses show that the amount of carbonate of lime held in solution in waters of the Mammoth Hot Springs is greatly in excess of that which the carbonic acid of the water is capable of dissolving. In the Cleopatra water, which contains the greatest amount of carbonate of lime, viz, 0.6254 parts in 1,000, the excess of carbonic acid over that necessary to form the neutral carbonate is 0.3537 gramme per litre. If this were united to form bicarbonates, the excess of free carbon dioxide would be but 0.0786 gramme. But as water saturated with carbonic acid, that is, containing 2 grammes per kilogramme, will dissolve but 0.88 gramme carbonate of lime, the proportionate amount dissolved by 0.3537 gramme of carbonic acid will be 0.1552 gramme of carbonate of lime. Since the water actually contains 0.6254 gramme of carbonate of lime in solution in each kilogram of water, there is an excess of 0.4698 gramme of carbonate of lime which has been dissolved either by increased pressure or by the alkaline salts present. As the water has been under pressure,

[1] Original Researches, p. 63.
[2] Carlsbad, Marienbad, etc., u. ihre Umgebung : Prag 1862, p. 76.

which was relieved as it rose to the surface, this has probably influenced the solution of the carbonate of lime, but the effect of the salts present is undoubtedly very important.

The Hierapolis waters contain 0.937 gramme of carbonate of lime per kilogram, considerably more than the Cleopatra water, with 0.3520 gramme of carbonic acid which can dissolve but 0.1549 gramme of carbonate of lime, leaving 0.7820 gramme of the latter to be dissolved by the 0.772 gramme of magnesium and sodium sulphates present, combined with the increased pressure under which the water existed before reaching the surface.

The Cleopatra water is supersaturated as it issues from the spring, since it deposits a small amount of calcic carbonate upon standing in tightly stoppered bottles. This supersaturation is probably due to the relief of pressure as the water rises in the tube of the spring and issues from the vent. As the water flows over the travertine slopes and basins there is a loss of carbonic acid and a deposition of carbonate of lime. At the same time the water is concentrated by evaporation owing to the large surface exposed. A small sample of water was collected from the slopes of the mound of the Cleopatra spring, at a point 25 feet below and distant 50 feet horizontally from the point of issue. The water, in flowing this distance, had cooled from 156° F. down to 113° F., and had lost 4 per cent by evaporation. This result is reached by a comparison of the sulphuric acid found in the water of the spring with that in the water of the slope. Correcting for evaporation, a comparison of the analysis with that of the spring water shows a loss of 0.2251 gramme per kilogram of carbonic acid by diffusion, and the deposition of 0.1675 gramme of carbonate of lime in flowing this short distance. Notwithstanding the deposition of this amount of calcium carbonate, the water was supersaturated with that salt, for it deposited carbonate of lime upon standing in a tightly stoppered bottle, probably because of the loss of the carbonic acid and the concentration of the water in flowing down the slope.

DEPOSITION OF CARBONATE OF LIME.

As the presence of carbonic acid gas is essential to the permanence of a solution of carbonate of lime, whether the solution contains alkaline and earthy salts or not, the withdrawal of the carbonic acid will cause a supersaturation of the liquid with a gradual separation and precipitation of the lime carbonate. Thus deposits of carbonate of lime may be due to the following causes:

   (1) Relief of pressure.
   (2) Diffusion of the carbonic acid by exposure to the atmosphere.
   (3) Evaporation.
   (4) Heating.
   (5) Influence of plant life.

Where the solution has been formed under pressure, the increased amount of carbonic acid which the water is then capable of retaining permits the solution of a larger amount of carbonate of lime. Upon the relief of this pressure the excess of gas escapes, and an unstable, supersaturated solution results, from which the lime carbonate gradually separates out. In this way originates the troublesome incrustations inside the pipes of pumps, and the saturation of many spring waters is undoubtedly due to the relief of pressure as the water issues from the ground.

Richly carbonated waters lose a portion of their carbonic acid upon exposure to the air; simple standing is sufficient to cause the separation of lime carbonate upon the surface of pools of such solutions, and the diffusion of the carbonic acid is proportionate to the temperature. Deposits formed in this way are common on the bottom of stagnant basins at the Mammoth Hot Springs, where the pellicle of carbonate of lime forming upon the surface breaks up on thickening, and falling to the bottom accumulates as a flaky, loose deposit. This diffusion of carbonic acid gas by exposure to the air is greatly facilitated by increasing the surface exposed, as well as by the agitation of the water; this is the case where the water spreads out over a surface in a thin sheet, or in cascades and spray. This diffusion is generally accompanied in such cases by evaporation, which also produces a separation of the lime carbonate. Stalactites, and similar formations common in limestone caves, are produced by these causes acting simultaneously, and the "petrified" or really incrusted bouquets, baskets, etc., of Carlsbad and many European springs, and also the Mammoth Hot Springs, are covered with crystals of calcite deposited in this way. (Fig. 53.)

Evaporation alone causes the formation of deposits of lime carbonate, in the form of *tufa*, by a concentration and supersaturation of the water. Such deposits are of great extent about several of the lakes of the Great Basin, as described by King, and Hague,[1] and lately by I. C. Russell.[2]

Heat causes a precipitation of the lime carbonate by a double action, driving off the carbonic acid and diminishing the solvent effect of the alkaline and earthy salts present,[3] resulting in the formation of boiler scale and incrustations where lime-bearing water is used for generating steam.

Deposits of carbonate of lime are also formed from natural waters by chemical reaction, as in the case of the tufa cones and tubes formed about the sublacustrine springs of Mono Lake.[4]

---

[1] Geol. Explor. of the 40th Par.: vol. 1, p. 514; vol. 2, p. 822.
[2] Lake Lahontan: Monograph No. 11, U. S. Geol. Survey.
[3] Skey : Trans. New Zealand Inst., vol. 10, p. 449.
[4] Russell, loc. cit.

DEPOSITS OF CARBONATE OF LIME DUE TO PLANT LIFE.

In the formation of the deposits just discussed, it is evident that plant life takes no part. It has, however, been long known that many water plants possess the power of abstracting carbonate of lime even from waters exceedingly poor in this salt, as in the case of sea water, where the Corrallines and other marine algæ build their framework of lime carbonate. Many fresh water plants, especially the *Charæ* and some mosses, also produce a separation of carbonate of lime. Our knowledge of this subject is chiefly due to the researches of Dr. Ferd. Cohn, who has shown the life of mosses and of algæ to be a most energetic factor in the formation of deposits of travertine.

The warm mineral waters of Carlsbad contain an abundant algous vegetation which forms thick cushions of jelly on the sides of the stream channels and generally wherever the warm waters flow. The association of this growth with the deposition of travertine is very striking, and early writers upon the vegetation of the springs called certain species lime-incrusted. Dr. Cohn was the first to discover the true relation of this plant life to travertine deposition, and in a paper published in 1862 he showed that these algæ actually eliminate carbonate of lime from the water and form travertine.[1] In proof of this he states that if a part of the algous jelly be pressed between the fingers an extremely fine sand is felt between the tips of the fingers, the grains being much larger if the jelly is taken from the older and lower parts of the growth. The nature of this sand and its true relation to the vegetable tissue are easily recognizable, the microscope showing minute crystals of carbonate of lime in the slime between the vegetable threads and upon their surface. These crystals, which at first are separate, increase in number and form star-like clusters, which by enlargement grow into grains of calcareous sand. By the further deposition of carbonate of lime these grains grow together and are cemented into solid travertine. All these steps are said to be recognizable under the microscope with the aid of hydrochloric acid.

The explanation of this deposition of lime carbonate within and upon the vegetable tissue is said to be the physiological action of the plant, which by withdrawing carbonic acid from the water diminishes the amount of carbonate of lime which it is capable of retaining in solution, the excess crystallizing out in the manner described. The supply of the soluble bicarbonate of lime withdrawn from the water by this double action of the plant is renewed by endomatic circulation. The cementing together of the grains of sand, which takes place in the older and deeper layers of the algous mass, is

[1] Die Algen des Karlsbader Sprudels, mit Rücksicht auf die Bildung des Sprudel sinters: Abhandl. der Schles. Gesell. pt. 2 Nat., 1862, p. 35.

thought to be largely due to a process independent of plant life, in which the porosity of the tufa plays a part.

The exact relation of the crystals and grains of carbonate of lime varies in the different species of algæ. In the *Oscillariæ* of Carlsbad, and allied species, the crystals form in the slimy inter-cellular tissue; in *Halimeda*, the carbonate of lime forms a sieve-like cover about the tips of the algæ filaments, and in *Acetularia* it occurs as a tube about the stalk of the plant. In the *Chara* the lime is separated and deposited in the cells and cell walls of the back alone, while in the *Corallines* it is found only within the cells.

It has already been stated that the algous vegetation of the Mammoth Hot Springs also produces a separation of carbonate of lime. A careful study of this vegetation in the field and under the microscope shows that this process is similar to that discovered by Dr. Cohn. At Carlsbad it was found that no vegetation was present and no tufa was deposited where the temperature exceeded 131° F. (44° R.), but with the appearance of algæ in the water the deposition of travertine began. A similar statement can not be offered in proof of the influence of such growths in the deposition of carbonate of lime at the Mammoth Hot Springs, for one species of algæ is found at 165° F., the hottest water of this locality. But that the algæ of these springs secrete carbonate of lime and form travertine can be satisfactorily demonstrated, and the process may be observed wherever the hot waters flow.

Masses of gelatinous vegetable growth, closely resembling those of the Carlsbad sprudels, are found about many of the springs, notably those at the north base of Capitol Hill and at the Jupiter Springs. In this vegetable jelly thin and flaky layers of carbonate of lime are found in the plant tissue. An examination of this gelatinous substance shows that it is composed of successive membrane-like sheets, in which minute gritty particles can be felt with the fingers. Under the microscope isolated little crystals and stellate accretions of these crystals are found scattered about in the plant tissue. These by further growth form minute grains of carbonate of lime. In the older layers and on the surface of this flaky travertine all sizes of grains are found, the largest being one millimeter in diameter.

This deposit is made up of these pellets, plainly seen in the freshly formed tufa layers, but indistinguishable in the older layers, where the grains are cemented together and the oolitic structure is lost. This cementation of these pellets and of the thin laminæ forms a firm, more or less compact, travertine. The membranous structure of the Carlsbad growth is supposed to be due to the intermediate or intercalated layers of carbonate of lime, ascribed to a certain periodicity of outflow from the spring, the temperature being constant. The same structure found at the Mammoth Hot Springs is not necessarily due to this cause, since alterations in the amount and temperature of the current nourishing the algæ may be caused by the obstructive growth of the plants themselves, which thus produces a change in both the vegetation and the deposit. In addition to this, evaporation and loss of carbonic acid from the water flowing over the surface of the growth cause the formation of a crust of carbonate of lime, which is afterwards covered by algæ, as the water dammed by the growth below is forced to flow over this surface.

If the water supply be cut off from a mass of such algous growth the plants die, the green changes to brown, and this to rose pink, and finally to a light salmon, while the odor of decaying vegetable matter is very perceptible. In a short time all color fades from the surface and a soft and porous chalky deposit is all that remains of the mass of jelly-like algous growth. If a little moisture is present the pink tint remains a long time, and is generally noticeable in the inner parts of the deposit. Such areas are common at the Mammoth Hot Springs, especially about the changing vents of sulphur springs, and at numerous places where warm waters have issued from little vents early in the season, but have dried up on the advance of summer, leaving this pink tinted and soft deposit as the only evidence of the recent outflow.

As already mentioned, the escape of carbonic acid and the evaporation of the water are very rapid on the overflow slopes, below the Main Springs. In consequence of this, a coating of pure white crystalline carbonate of lime is rapidly deposited upon objects of any kind placed in the spray of the hot water, a thickness of $\frac{1}{26}$ to $\frac{1}{18}$ of an inch being formed in three days under favorable circumstances. This incrusting property of the water is utilized for the production of coated "specimens," made to sell to tourists. Horseshoes, pinecones, bottles, and different forms of wire work are placed on rude racks, or suspended from the cross-bars of the rack by strings and the hot water led over them so that the objects to be coated are constantly wetted by the hot spray, as shown in Fig. 53. The deposit so formed is pure white and marble-like, and the little crystals sparkle in the light. If, however, the objects be permitted to remain in the spray several days longer, the deposit loses its intense whiteness and assumes a dull yellowish tint. At the same time the former smooth surface of the coating is dotted with wart-like excrescences which become larger and more numerous the longer the specimen is exposed, and in time will distort and disguise the original shape of the object, while the color becomes a rich umber brown. By treating specimens of this character with dilute hydrochloric acid these changes are seen to be due to the growth of algæ. The first points of growth are places of most rapid deposition, and warty excrescences are formed; later the algæ are present all over the surface and the thick coating becomes dendritic in structure, and both color and form suggest organic life. In such deposits vegetable life is not, however, the only factor, since we have seen that the water will deposit a coating of carbonate of lime without the influence of plant-life. But these influences are eliminated if we place such objects under water, in the bowl of either of the main springs; for bottles, horseshoes, or other articles of glass or iron immersed in these springs for a long time were not coated.

If, however, a pine branch, a part of some bush or plant, be placed under the water it is shortly covered with an incrusting cylinder of carbonate of lime. The surface of this cylinder is reddish brown and warty, resembling the deposit last mentioned; the interior is dendritic in structure and of a light buff color. This deposit closely resembles the travertine cylinders formed about twigs and branches at Tivoli. In the formation of these cylinders at Tivoli, Cohn has shown[1] that crystalline sinter is only separated about living plants whose bark is covered with growing algæ and mosses. In our deposits the nature of the substance immersed is important only because it affects the ability of the algæ to obtain a foothold and to grow, and the deposition of sinter is coincident with the growth of such algæ.

A portion of this deposit dissolved in dilute hydrochloric acid leaves a residue of tangled algous filaments, forming a felt-like mass. The nodular masses common on the bottom of hot water pools and basins are similar in nature; the surface of these formations is moss-like, brown and greenish in color, particularly in the depressed portions. The interior is formed of buff-colored tufa of radiating dendritic fibers.

DESCRIPTION OF THE DEPOSITS.

The different varieties of travertine found at the Mammoth Hot Springs vary in physical structure and in appearance, according to the conditions under which the deposit was formed. In general, the compactness of the travertine depends upon the rapidity of formation, some of the most quickly formed deposit being so light and porous that it is easily crumbled to powder between the fingers, while the slowest formed deposit is almost flint-like in texture.

The travertine of the older terraces is often compact, dense, and hard, resembling an ordinary limestone; another variety, often of recent formation, is also compact and crystalline, resembling the purest of marble; while the freshly formed walls of the basins of the Main Terrace are often soft and easily crushed. Notwithstanding the marked differences of structure and appearance the travertine all has the same chemical composition as shown by analysis.

The following analyses, made for the Geological Survey of the

[1] Neues Jahrbuch (Leonhard), 1864, p. 580.

Yellowstone National Park, by Mr. J. E. Whitfield, show the composition of typical specimens of varying forms of the Mammoth Hot Springs deposit; analyses are also given of travertines from other localities.

Analyses of Travertines.

	I.	II.	III.	IV.	V.	VI.	VII.	VIII.	IX.
SiO_2, silica	0.08	0.26	0.06	0.01	0.15		0.6	0.30	0.12
$Al_2O_3 + Fe_2O_3$	0.15	0.11	0.14	0.05	0.49			1.10	
SO_3, sulph. acid	1.72	1.34	0.70	0.49	0.55			0.80*	0.08*
CaO, lime	53.83	54.09	55.02	55.02	53.46				
$CaCO_3$, lime carbonate	*(94.97)	*(93.77)	*(96.02)	*(96.02)	*(93.36)	96.82	98.02	97.00	95.62
MgO, magnesia	0.90	0.60	0.06	0.07	0.42			0.16ᵃ	3.06ᵃ
NaCl, sodium chloride	0.02	0.26	0.20	0.12	0.13				
K_2O, potash			0.08ᵃ	0.04	0.01				0.10*
CO_2, carbonic acid	41.79	42.14	42.25	42.25	41.96				
H_2O, water	1.43	1.19	1.05	1.61	2.44	1.50			
C, carbon	0.21	None	0.24	0.11	0.37				
Other constituents						1.41	1.20		
Total	100.13	99.96	99.81	99.77	99.98	99.94	100.00	99.36	99.39

*Sulphate of lime.   ᵃCarbonate of magnesia.   *Potassic chloride
*If all the CO_2 be supposed to be combined with lime.

No. I is a compact yellowish travertine from the slopes below the Hotel Terrace, and it represents the older travertine.

No. II is the riffled travertine forming the ridge west of the Blue Springs and above Cupid's Cave.

No. III is the fibrous white travertine forming the fan-shaped masses seen in the Blue Springs and elsewhere, the specimen being from a deserted vent near the Blue Springs.

No. IV is from a mushroom-shaped mass, showing the color and structure of the organic growth found in the overflow of spring No. 24.

No. V is the crystalline travertine found on the walls of Cupid's Cave. The surface is satiny and mottled, with spicules and beaded formations resembling siliceous sinter.

No. VI is the analysis of the Carlsbad sprudelstein made by Berzelius.[1]

No. VII and No. VIII are travertines from Hierapolis and from Kukurtlu, Asia Minor, analyzed by J. Laurence Smith.[2]

No. IX is the tufa found about the Arkansas hot springs, analyzed by David Dale Owen.[3]

Though travertine formed without the presence and aid of plant-life forms but a very small part of the bulk of the Mammoth Hot

[1] Annalen der Physik, Gilbert: vol. 74, p. 168.
[2] Original Researches, p. 65.
[3] Geology of Arkansas.

Springs deposit, there are two interesting varieties in the formation of which vegetable life was absent. The first is the thin flaky deposit found at the bottom of stagnant pools and basins of the spring water. This is formed by a separation of calcic carbonate at the surface of the pool, owing to the diffusion of carbon dioxide upon prolonged exposure of the water, forming a thin wax-like film upon the surface; this thickens until the crust breaks up from its own weight and the flakes settle to the bottom of the basin. This material is nearly pure carbonate of lime, whose specific gravity, 2.70356, shows it to be a true calcite.

Another variety also made independently of plant life is that which forms the lining of hot-spring chambers, such as the Devil's Kitchen, and the spring vent-holes. This is deposited comparatively slowly, and occurs in shelly layers half an inch to three inches thick, with a smooth, rounded, and globular surface. It is crystalline and marble-like and pure white. This travertine is a crystallization out of a supersaturated solution of carbonate of lime, due to the relief of pressure as the waters approach the surface. A similar deposit lines the vent-holes of the Orange and other springs, and is analogous to the deposits so quickly formed in the conduit pipes leading the hot water to the hotel baths, also due to supersaturation, experiments showing that such solutions do not deposit their excess of lime at once, but in the course of a short time.

All other varieties of the travertine so far found have been formed partially or entirely by the aid of plant life. Of the numerous forms produced in this way there is none which shows its vegetable origin more clearly than the fibrous tufa forming the fan-like masses found in many of the springs. (Fig. 52.) A simple examination of this deposit with a lens shows that the fibers are neither long crystals nor crystal aggregates, and the stringy or blade-like fibers suggest the incrustation of vegetable filaments. The upper surface of this deposit is fairly even and the fibers round and parallel in arrangement. The inner part of the specimen is similar, but the fibers are sharper and resemble blades of grass arranged loosely, giving an open and porous structure. The under surface of these fans is more uneven, the fibers are round and covered with little pellets of lime sometimes clustered in botryoidal forms, while the threads themselves are irregularly arranged, as if a skein of silk floating in the shifting currents of a stream were suddenly turned to snowy travertine. Plate LXXXI shows the upper surface of a part of one of these travertine fans, on which the travertine frost-work is particularly beautiful. The specimen is formed of a stringy or fibrous deposit covered by gnarled and knotted ropy forms, whose surface is covered by aggregated travertine pellets of varying sizes up to one-eighth of an inch in diameter, these in turn coated with a drusy frost-work of little crystals. Bubble-like shells of translucent wax-

like travertine, sometimes entire, oftener broken, lie between the
fibers or entangled in the network, their broken edges beaded and
their surface dotted with minute pellets of the same material. An
examination of a fragment of this porous fibrous travertine with a
pocket lens shows that the fibers, tubes, and blades are built up of
minute rods lying alongside of one another and cemented into bun-
dles or plates. Each rod has a hard vitreous center, with an outer
more opaque coating. Dissolving a little fragment of the travertine in dilute hydrochloric acid shows that each little rod is formed
by a single algæ thread. Remembering the occurrence of the fans
of this fibrous travertine, we can only conclude that the formation
is produced by the white species of algæ so common in the hotter
waters of the springs.

The curious mushroom-shaped forms found in the channels of
many of the springs are detached with difficulty, as the deposit is
quite hard when fresh. The top is usually wet by the ripples and
spray of the stream, but is above the general body of water. This
upper surface is riffled by a network of little ridges one-eighth of
an inch high, with basin-like depressions between. The color is a
bright orange red, which is most brilliant in the depressions, where
one familiar with algæ at once recognizes the vegetable nature of
the color. A transverse section of the specimen proves that it is
also of algous origin. The stem consists of fibrous travertine resembling that forming the "fans." This forms the center or middle
layer of the cap of the toadstool also, but is overlaid by a layer one-
quarter to three-quarters of an inch thick of quite different structure.
This top layer is also fibrous, but the fibers are short, stout, and
perpendicular to the underlying deposit. The under side of the cap
is coated with hard, porcelanous travertine, with smooth surface,
often dotted with botryoidal clusters of white pellets, to which
sulphur-coated filaments are often attached. Both varieties of
fibrous travertine show a netted mass of algæ filaments when dissolved in dilute acid. The most common variety of travertine,
forming the riffled surfaces of the rounded slopes, benches, and terraced basins, is like that forming the top layer of the mushroom
forms just mentioned.

The riffled surface is due to innumerable little ridges which run
across the surface in wavy lines, and, meeting, form miniatures of the
larger basins. If the slope is very gentle these little basins are proportionately larger and the dividing walls very thin, while on steeper
slopes the ridges are thick and close together, producing a reticulated
surface. While wet by the hot water the color is generally quite
brilliant. If the volume of water be large and the current rapid,
the color is a creamy white, shading at the shallower and less rapid
parts of the overflow into pale salmon and pink, and these to orange,
red, and burnt sienna. If this deposit be examined with a lens the

TRAVERTINE, MAMMOTH HOT SPRINGS.

color is seen to be due to a fuzzy growth of algæ, and if a fragment
be carefully dissolved in dilute hydrochloride the fuzzy coating is
found to be only the tips of the living ends of algæ threads buried in
the deposit beneath. The structure is fibrous and quite like the upper
layer of the form last described. Where the travertine forming the
riffled slopes is broken down, showing its general structure, it is seen
to consist of concentric shells or curved plates of varying density
and thickness. This evidences varying conditions of deposition,
such as changes in the supply, and consequently of temperature,
affecting the nature of the plant growth.

Changes of structure are easily produced in this way, as the compactness of the deposit depends largely upon the rapidity of deposition, being least where most quickly formed. Changes may also be
due to the effect of sudden cold or the different seasons, as intense
cold might kill the plant growth, and the less evaporation of the
winter months, with a probable less vigorous growth of the algæ,
would produce a thinner, more compact layer of sinter. In general it
may be stated that variations in evaporation and in the growth of
the algous vegetation will produce variations in the structure of the
deposit.

Another common variety is formed of overlapping layers of fibrous
travertine, resembling a thatched roof; it is but another form of
that making the fan-shaped masses and is produced by either the
white or green filamentary algæ. This deposit originates the pillars
of the Pulpit Basins and of others like them.

At the edges of the Main Springs is a very hard laminated sinter
formed by the evaporation of the water but tinted by the plant life
present. The lining of these bowls and of the adjoining pools is
formed of a mossy deposit already described.

Coralloidal travertine is found in many quiet basins and pools
where the water is concentrated by evaporation and the lime crystallizes out upon the web of algæ threads present in the water, producing very delicate forms resembling certain species of corals. Sometimes such deposits support the pellicle of lime gathering on the
surface, and thus the pool is completely roofed over. The stems of
this variety of tufa are thickly set with a drusy coating of crystals
arranged perpendicularly to the surface of the stem. The honey-
combed deposit found in many of the dry and empty basins is formed
by the rising of gas bubbles through the soft, gelatinous mass of the
algæ, the tubes remaining open during the conversion of the growth
into solid travertine.

WEATHERING OF THE TRAVERTINE.

Deprived of their supply of water, the travertine slopes lose their
brilliant colors, which soon fade out, leaving a chalky white surface;
this darkens by prolonged exposure to a light gray and in a few

years to the dull gray tone of the older deposits. This gray tint is only a surface coloration, for the deposit beneath is still pure white.

Frost is the greatest foe to the preservation of the basins and terraces. In winter the cool overflow from the springs, with rain and melted snow, freezes upon the surface of the deposit, and thickening, tears off the walls of the terraced basins by its weight, or freezing in the porous travertine and in its cracks and fissures, opened by the settling of the deposit, pries off and loosens many of the most beautiful forms of the tufa. A judicious distribution of the overflow from existing springs would, however, rebuild and repair many of the ruined and crumbling slopes and basins without detracting from the beauty of other parts of the deposit.

Infiltrating waters from the overflow of the springs carrying carbonate of lime, effect a change in the open and porous tufa, hardening it into a denser and more compact rock. The travertine is also altered by steam and sulphurous vapors rising through it; steam alone often produces a coarsely granular structure of loosely compacted crystals. Where the vapors are sulphurous the tufa is converted into acicular crystals of gypsum, generally preserving the open structure of the travertine, with sulphur deposited in the open spaces.

ORIGIN OF SILICEOUS SINTER.

With the exception of the calcareous waters of the Mammoth Hot Springs already described, and a few less important localities, the hot waters of the Yellowstone National Park, like those of other volcanic areas, are characterized by the proportionately large amount of silica contained in solution. These springs, like those of Iceland, may be divided into two groups, of acid siliceous and of alkaline siliceous waters—a distinction quite sufficient for the purpose of this article. The acid waters include the Highland springs, those of Crater Hills and several other localities in the Park, and are generally characterized by deposits of sulphur and efflorescent alum salts, while the waters contain free hydrochloric or sulphuric acids. The alkaline springs form the largest of the two groups, comprising the geysers and the other hot springs of the Geyser Basins, and similar hot-spring areas.

About these alkaline hot springs the mineral deposits consist almost entirely of silica, partly as opal in the clay and less decomposed rhyolite, but chiefly as siliceous sinter, a surface incrustation of white amorphous silica. This sinter forms the mounds and cones of the geysers and springs, the fretted and scalloped rims of the quieter pools, and the great white flats surrounding the springs.

Although such deposits of siliceous sinter are found wherever geyser action is manifested, and quite commonly in connection with alkaline hot springs in all parts of the world, the deposits of Iceland, New Zealand, and the Yellowstone Park are much the best known and far exceed those of other localities. The Iceland deposits have been known the longest, and have been studied by many observers. The Haukadal area is the most familiar, as it is here that the Great Geyser is situated. At this place the white sinter deposits cover many acres of ground and form the snow-white basins of the quietly boiling springs, and the mounds of Strokr and the Geyser.

The New Zealand sinter areas are similar in character, but the deposits are much more extensive than those of Iceland. In many parts of the North Island there are sinter flats and mounds resembling the Iceland and Yellowstone deposits, but in neither of these countries is there anything to equal in beauty the wonderful stalactic basins of the pink and white terraces of Lake Rotomahana, which were destroyed in the volcanic outbreak of 1886. The sinter deposits of the Yellowstone National Park are, however, the largest known, covering many square miles at the different geyser basins and other hot spring localities. There is probably no better field in which to study the different varieties of silica deposited by hot spring waters and to observe the conditions under which they are formed. In the series of observations carried on by the writer, it has been found that a large proportion of the siliceous sinter of the different geyser basins is formed by the agency of vegetable life, algæ and mosses living in the siliceous water, and it is these deposits and their origin that are of special interest in this part of the present paper.

The formation of siliceous sinter by plant life has been found to be going on at many hot-spring areas in the Park, in fact wherever alkaline siliceous waters are found; but since it is necessary to select some particular place where the details of such growths may be described, the Upper Geyser Basin of the Firehole River is chosen, as it is easily accessible and is seen by all visitors to the Park.

UPPER GEYSER BASIN OF THE FIREHOLE RIVER.

GENERAL DESCRIPTION.

The Upper Geyser Basin of the Firehole River lies 10 miles west of Yellowstone Lake, 39 miles south of the north boundary, and 11 miles east of the western boundary of the Park, and is reached by a stage ride of 8 miles from the Lower Firehole, or of 50 miles from the Mammoth Hot Springs. The altitude is about 7,300 feet above sea level, but though the basin is quite near the continental divide which separates the drainage of the Pacific from that of the Atlantic Ocean, it occupies a well-marked depression in the great rhyolite plateau of the Park. Situated almost on the crest of the continent, it is yet somewhat distant from the mountain ranges of the Park, and is about equally removed from the Gallatin range on the north, the volcanic peaks of the Absarokas on the east, and the Teton uplift on the south.

The area of about two square miles comprising the Upper Basin, as it is commonly called, has a fairly level surface, 1½ miles in extreme width, and 2¼ miles long. This is inclosed between the abrupt cliffs of the Madison Plateau on the west and north and the heavily wooded slopes of the continental divide on the south and east. The rhyolite which forms these barrier walls, and is well exposed in the cliffs about the basin, also forms the floor of the valley itself, but is generally concealed either by a sheet of siliceous sinter in the vicinity of the hot springs, or by its own débris, a black pearlitic sand. The Firehole River and its branches, the Little Firehole and Iron Creek, drain the basin ; the Firehole flows through the eastern part of the area, and most of the geysers and hot springs are situated upon or close to its banks. Except where covered by sinter or hot water marsh, the surface is timbered with a thick growth of pine (*Pinus Murrayana*), and scattering trees are also found over the older, disintegrating deposits of sinter. These sinter areas form the most

Fig. 54. General view of part of the Upper Geyser Basin

striking feature of the topography of the basin, and the bare white flats and sinter mounds are in strong contrast with the dark green of the neighboring forest. The mass of sinter deposited by the hot water is so large that it has materially changed the original surface of the ground. Fig. 54 shows the appearance of the central part of the main geyser area, seen from the slopes near the Grand Geyser.

This sinter covering is variable in thickness, the maximum depth being about thirty feet around several of the older vents, a thickness which attests the great age of the thermal action. The sinter sheet is constantly extending its boundaries, as shown by dying and dead trees and other vegetation standing in the silica, but a gradual dying out of the older vents permits a slight surface disintegration of the deposit, with a gradual encroachment of vegetation upon it.

But few of the many hundred springs of the Upper Basin are turbid or muddy, and the pools are generally characterized by the exquisite clearness of the water, which appears of varying shades of blue or green, according to the depth and amount of light admitted. If the water be quiet, the transparency is such that the minute details of the basin can be seen, even at depths of 20 feet or more. In the quieter springs, where the temperature does not exceed 150° F., the basins are often lined with a more or less abundant algous growth, whose orange, red, yellow, brown, and green tints impart new shades to the water. Though there are many of these *laugs* the greater number of springs possess a temperature approaching or equaling 198° F., the boiling point at this altitude, and in such springs the water is usually in constant or intermittent ebullition. Around the margins of such springs a rim of silica is generally built up by the hot waters. The inner surface, where constantly wet by steam and spray, is a bright, tawny yellow, the deposit being sponge-like in form and in color, though composed of hard silica. The outer surface of these rims is generally gray, often ornamented with pearls or beadwork of most delicate structure.

At the margins of more tranquil springs a flat projecting crust or edging of white sinter is found, sometimes extending out over the water and even roofing over parts of the spring, as shown in Fig. 55. This edging of white sinter is oftentimes scalloped in outline, each scallop closely resembling fungus growths common on the bark of trees in damp woods. Many of the springs have received appropriate names, such as Sapphire Pool, the Morning Glory, and Chromatic Spring, but many more, perhaps equally beautiful, remain unnamed. It is not easy to draw a sharp line between hot springs and geysers, nor is it at all necessary ; for there is every gradation, from a quiet pool with simple intermittent increase in temperature to the great fountains of boiling water which provoke our wonder and our admiration. Forty-eight geysers are known in this basin each possessing such peculiarities of eruption or surroundings as to make it of interest and distinguish it from its fellows. Most of these were named by the earlier visitors to the Park, who christened the Giant, Bee Hive, Old Faithful, and other equally familiar geysers.

Even this brief description of the Upper Geyser Basin is incomplete without some mention of the brilliant colors noticed wherever the hot waters flow. These multitudinous tints of red and yellow, green and brown, are all produced by the growth of hot-water algæ, which, as I shall show further on, eliminate silica from the hot waters by their vital growth, and contribute largely to the building up of the sinter deposits, besides giving them their brilliant tints.

Fig. 55. Avoca Spring, Upper Geyser Basin.

CHARACTER OF THE HOT SPRING WATERS.

The hot waters of the Upper Basin are mostly clear and perfectly transparent and show in perfection the blue-green tints of pure water in the many spring bowls and basins. In many of the springs an iridescent effect is seen in the water, which is not due to a film on the surface, but is caused by the reflection of light from circulating currents. Tested with litmus paper, the water is either neutral or feebly alkaline in reaction, and it generally possesses a slight sulphurous odor. When cold it is flat and insipid in taste and scarcely palatable, but is not injurious. These alkaline-siliceous waters are similar in character but vary slightly in the amount of material held in solution. Chemical analysis shows this to consist of silica and of readily soluble alkaline and earthy salts, which are retained in solution and carried off by the surface drainage, while a large part of the silica is deposited about the springs and geysers.

The following analyses, made for the Geological Survey of the Park by Prof. F. A. Gooch and J. E. Whitfield, show the composition of the geyser waters of this basin. Analyses are also given of the water from the Great Geyser of Iceland, and from the New Zealand geysers, the former by Damour,[1] the latter by Smith.[2]

Analyses of geyser waters.

[Constituents grouped in probable combination. Grammes per kilogram.]

	Asta Spring.	Splendid Geyser.	Grand Geyser.	Old Faithful Geyser.	Great Geyser, Iceland.	White Terrace Geyser, New Zealand.
SiO_2, silica	.1650	.2964	.3035	.2961	.5193	.5060
NaCl, sodium chloride	.1320	.4940	.5643	.6393	.2379	1.0220*
LiCl, lithium chloride	.0048	.0140	.0218	.0040		*.0350
KCl, potassium chloride	.0241	.0231	.0319	.0478		
KBr, potassium bromide			Trace	.0051		
Na_2SO_4, sodium sulphate	.0575	.0281	.6387	.0270	.1342	
$Na_2B_4O_7$, sodium borate		.0635	.6350	.0213		
Na_2AsO_4, sodium arseniate		.0025	.0014	.0027		
Na_2SiO_3, sodium silicate				.0279		1.2230
Na_2CO_3, sodium carbonate	.1463	.3286	.3209	.2088	.4567	
$MgCO_3$, magnesium carbonate	.0065	.0026	None	.0021		Trace.
$CaCO_3$, lime carbonate	.0295	.0075	.0070	.0038		.025
$FeCO_3$, iron carbonate		.0001		Trace		
Al_2O_3, alumina	.0212	.0051	.0061	.0017		.005
H_2S, hydrogen sulphide			Trace	.0002		
NH_4Cl, ammonium chloride		.0002	.0012	Trace		
CO_2, carbonic acid	.1045	.1980	.0387			
K_2SO_4, potassium sulphate					.0180	.0750
$MgSO_4$, magnesium sulphate					.0091	
Na_2S, sodium sulphide					.0088	
Total	.6764	1.6340	1.3905	1.3908	1.2935	2.6570
Specific gravity		1.00132	1.00108	1.00096	1.00095	1.00067

*$CaCl_2$ †Na_4O.

FORMATION OF SILICEOUS SINTER.

The separation and deposition of silica from hot spring waters in the form of siliceous sinter has been ascribed by different writers to one or more of the following causes :

(1) Relief of pressure.
(2) Cooling.
(3) Chemical reaction.
(4) Evaporation.

At the Norris Geyser Basin the first two causes produce a separation of silica from the hot waters, but the waters of the other geyser basins contain very much less silica, and as far as observed neither relief of pressure nor cooling will produce a separation of the silica. Water collected from the springs and geysers of the Upper and Lower Geyser basins was perfectly transparent, and remains clear

[1] Ann. Chem. u. Pharm., vol. 62. 1847, p. 49.
[2] Jour. für prakt. Chemie, vol. 89, 1863, p. 186.

and without sediment after standing several years. Experiments in
the laboratory show that the silica in these waters remains dissolved
even when the water is cooled down to the freezing point, and it is
only after the crystallization of the water by freezing that the silica
is separated and settled down as an insoluble flocculent precipitate
upon melting the ice.

The formation of sinter by the waters of the Iceland geyser, which
analysis shows to be similar to the waters of the Upper Basin in
character, but more heavily charged with silica, is explained by
Damour[1] and Descloiseaux by supposing the silica to be present in
solution, as an alkaline silicate, which is decomposed by uprising
sulphurous and hydrochloric vapors into free hydrated silica and
alkaline salts. From the supersaturated solution of silica, formed
in this way the silica separates out in the form of sinter. In several
analyses, Damour found a constant relation of 3:1 between the oxygen of the silica and that of the bases; when the alkalies are present
partly as chlorides and sulphates, formed by the decomposition of the
alkaline silicates, the relation existing between the oxygen of the
silica and that of the bases of the undecomposed silicates was found
to vary from 1:5 to 1:9, and wherever the latter proportion pre-
vailed, as it does in the water of the Great Geyser, silica is deposited, the amount deposited each day corresponding to the quantity of
the alkali saturated in that time by the action of the acid vapors or
by the oxidation into sulphates of the alkaline sulphides in contact
with the air. Laboratory experiments show that the waters of the
Upper Basin remain unaltered upon saturating them with hydrogen
sulphide, and that the silica is probably present in solution as free
hydrated silica.

LeConte and Rising[2] suppose the precipitation of silica taking
place at Sulphur Bank, Cal., to be due to the neutralization of the
upcoming hot alkaline waters by descending acid solutions, a process evidently not in operation at the geysers of the Upper Basin.

Roscoe and Schorlemmer[3] state that the alkaline silicates of the
Iceland waters are decomposed by the carbonic acid of the atmosphere with a formation of alkaline carbonates and free silica, the
latter being deposited. This gas, passed through the Upper Basin
waters for several hours, produced no visible effect upon the water.
Bunsen, to whom we are indebted for the accepted theory of geyser
action, ascribed the formation of the Iceland sinters to the evaporation of the water.[4] His experiments showed that the silica of the
geyser water was not deposited upon cooling and only separated out
upon the advanced concentration of the water, but was readily de-

[1] Philos. Mag., London, 1847, vol. 30, p. 405.
[2] Am. Jour. Sci., 3d series, vol. 24, p. 33.
[3] Treatise on Chemistry, vol. 1, p. 571.
[4] Pseudo-volcanic phen. of Iceland: Memoirs of Cav. Soc., Graham, 1848, p. 336.

ALGÆ CHANNELS, EMERALD SPRING, UPPER GEYSER BASIN.

posited by evaporation. This cause produces some of the siliceous sinter found about the hot springs and geysers of the Upper Basin, but it is to the vegetation present in these hot waters that we must credit the formation of the greater part of the siliceous deposits of the geyser basins.

ALGOUS VEGETATION OF THE HOT WATERS.

Algæ are found in the thermal waters of the Upper Geyser Basin wherever the temperature is not too high to permit their development. The limiting degree of heat at which they have been found is 185° F., but the algous filaments are often found at that temperature, though such plants are immature and poorly developed, and it is not until the waters have cooled down to a temperature approximating 140° F. that these growths attain their full development. In these cooler waters their vegetable nature is more easily recognizable, for the waving green filaments, or the red and brown leathery sheets lining the springs, closely resemble sea weeds found on our coasts. But in the hotter waters the material hardly suggests the presence of vegetable matter, the densely gelatinous substance resembling mineral or possible cartilaginous animal material. The colors of these growths are generally quite brilliant, either golden-yellow, orange, or red, and in the hottest waters pale flesh-pink, or even white. These algæ are often so thickly encrusted by silica that the plant structure is not recognizable even under the microscope, and their presence is often only to be distinguished by the color. It has been found that the color of the growth depends upon the temperature of the water so that differences in color mark different degrees of heat. Some of the most striking color effects of the Upper Basin are due to this fact, such as the ribbon-like stripes of overflow channels, and the concentric rings of color found in shallow flaring hot spring bowls, and reaching a wonderful development in the Prismatic Lake of the Midway Basin.

The general sequence of colors is well illustrated by the occurrence of such growths in overflow streams with a constant volume, such as the outlet of the Black Sand. As the water from this spring flows along its channel it is rapidly chilled by contact with the air and by evaporation, and is soon cool enough to permit the growth of the more rudimentary forms which live at the highest temperature. These appear first in skeins of delicate white filaments which gradually change to pale flesh-pink farther down stream. As the water becomes cooler this pink becomes deeper, and a bright orange, and closely adherent fuzzy growth, rarely filamentous, appears at the border of the stream, and finally replaces the first-mentioned forms. This merges into yellowish-green which shades into a rich emerald farther down, this being the common color of fresh-water algæ. In the quiet waters of the pools fed by this stream the algæ

9 GEOL——42

present a different development, forming leathery sheets of tough
gelatinous material with coralloid and vase-shaped forms rising to
the surface, and often filling up a large part of the pool. Sheets of
brown or green, kelpy or leathery, also line the basins of warm
springs whose temperature does not exceed 140° F., but in springs
having a higher temperature the only vegetation present forms a
velvety, golden-yellow fuzz upon the bottom and sides of the bowl.
This growth is rarely noticed in springs where the water exceeds
160°, except at the edge of the pool. If the basin is funnel-shaped,
like that illustrated in Fig. 56, with flaring or saucer-shaped ex-
pansion, algæ grow in the cooler and shallower water of the margin,
forming concentric rings of yellow, old gold and orange, shading
into salmon-red and crimson, and this to brown at the border of the
spring. Around such springs the growth at the margin often forms
a raised rim of spongy, stiff jelly, sometimes almost rubber-like in
consistency, and red or brown in color. Evaporation of the water
drawn up to the top of such rims leaves a thin film of silica, which
thickens to a crust and so aids in the production of a permanent
sinter rim.

Where the overflow from a spring spreads out over the surface of
a mound the algæ often grow in cushions of red, white, brown or
green jelly, generally mistaken for simple gelatinous silica colored
with iron, and indistinguishable from simple mineral material to the
naked eye, though the putrid odor of such material removed from
the water and allowed to decay indicates its organic nature. In the
pools and basins about most of the geysers the bright orange algæ
form a velvety nap upon the smooth surface of the sinter. This is
easily recognizable in the pools about the Jewel Geyser, but the
same growth occurring in the basins about Old Faithful, and very
generally in the overflow channels of all the geysers, is so obscured
by the silica deposit about it that it is only noticeable because of its
brilliant color and the slippery feeling it imparts to the surface.

ALGÆ POOLS AND CHANNELS.

The vegetation of the hot spring waters attains its maximum development in the self-formed pools and basins found near the Emerald and the Black Sand springs of the Upper Basin and the Jelly
Spring of the Lower Geyser Basin. This is largely because of the
great and constant volume of the overflow from such springs, taken
in the natural water-ways and pools which the algæ form, and distributed by a nicely-adjusted system, by which the continual increase and growth of different parts of the overflow area are pro-
moted and fostered. At such places the formation of sinter by these
plant growths goes on rapidly, and the various gradations may be
seen, from the soft jelly to the firm and hard sinter, into which it is
transformed.

ALGÆ BASIN, EMERALD SPRING, UPPER GEYSER BASIN.

U. S. GEOLOGICAL SURVEY

NINTH ANNUAL REPORT PL. LXXXII

One of the best places to observe such pools is the flat about the Emerald Spring, where the sinter is all of algous origin. The Emerald Spring, whose clear green depths make the name a most appropriate one, is situated on the west bank of Iron Creek, about three-fourths of a mile west of the Upper Geyser Hotel. The bowl is 36 feet wide, 30 feet to 40 feet long, and 35 feet deep, surrounded by a shallow basin or marginal area 1 to 5 feet wide, outlined and rimmed by a low border of firm algous jelly whose upper surface is whitened by silica left by evaporation. The water is apparently perfectly clear and free from suspended material, and it possesses a temperature of 150° in the spring bowl. The bottom and sides of this bowl are formed of a creamy gray siliceous mud, particles of which are probably held in suspension in the water. This is covered by a fuzzy growth of light canary-yellow algæ. This coloration is best seen near the edge of the bowl, though in the shallower water of the margin the color is deeper, and in the recesses where the temperature is but 135° F. the growth is brownish-green at the surface, underlaid by bright red, and forms a soft, leathery sheet on the bottom.

The overflow of the spring is very uniform in volume, and leaves the basin at the southwest end, running off in a channel 2 to 4 feet wide and half an inch to 2 inches deep. This channel is lined with a thin membraneous sheet, whose gamboge yellow shades gradually into a yellowish green, with dull red, and olive greens at the borders of the stream. Twenty feet from the outlet the stream broadens into an area of algæ channels and basins, outlined and dammed up by the algous growth. Pl. LXXXII shows a portion of this area, photographed after draining off the water. In the foreground is the water-way, inclosed by the algous growth at the sides, its floor dotted with insular masses of the same material, the normal water level being up to their tops. In the background is seen a part of the surrounding flat with dead tree trunks standing upright in the sinter, their lower portions white with silica left by the evaporation of water drawn up by capillary force. The algous forms seen in this illustration are quite characteristic of the growth where the conditions permit its full development. This water-way is floored with a sheet of olive or emerald green, kelpy jelly. Where there is a moderate current this lining is nearly smooth, resembling a sheet of wet leather, but in quieter water this soft carpet is dotted with warty excrescences and little pillars produced by their upward growth ; the latter sometimes terminates by balloon-like caps or globes containing a bubble of gas. When in the early stages of their growth these slender spines or pillars consist of soft gelatinous material, sinking to a shapeless mass of jelly when removed from the water, but as they increase in height and in diameter a firmer siliceous center is formed which gives stability to such shapes. When by their upward growth these pillars reach the surface of the pool they increase rapidly in diameter, particularly at the water level, and a cup-shaped cap or crown is soon formed upon the pillar, often with a vase-like shape. If several of these grow near together the caps extending laterally soon unite, and form the peculiar masses seen in Pl. LXXXII. The continued growth of new pillars gradually fills up parts of the channel and eventually pond back the water, partially at first and at last entirely. In this case the increased depth of water resulting permits a further upward growth of the algæ, and a series of pools or basins sometimes results, in which the water levels are quite different, while the water cooled in passing from pool to pool possesses different temperatures in each. A close view of such a basin is shown in Pl. LXXXIII, part of the same area shown in Pl. LXXXII. The algous growth has here dammed up a channel forming a little basin already partly filled with isolated pillars and aggregates these growths. The continued growth of the algæ raises the water of level until finally the enfeebled current brings but a small supply, with a consequent gradually lowered temperature, not only for the basin itself, but also in adjacent pools whose supply may have been • entirely cut off. In such cases, the nature of the growth changes; it is not known that life ceases, though it seems probable that these algæ die, and new species are introduced. At any rate, the bright-colored algous jelly forming the outer covering of the pillars and algæ vases changes to light salmon pink, and the substance itself becomes noticeably siliceous or forms a filmy web upon the siliceous center. There is as yet no increment of silica, but a simple shrinking and hardening of the gelatinous envelope, but if the temperature be gradually reduced to 85° F. or 90° F. these forms become coated with a mossy incrustation of hard silica, and the algous structure and outlines are obscured or concealed by the coral-like coating.

If instead of this gradual reduction of the volume and temperature of the supply, the water is completely shut off suddenly, the gelatinous material dries up, for the water in the basin either evaporates or oozes through the porous growth. In this case the algæ soon lose their bright colors, which fade like those of the Mammoth Hot Springs to a delicate salmon pink, and finally pure white, becoming light but firm structures of opaque white, hydrated silica. Generally the pools have been filled up by the pillars, and oftentimes completely roofed over by their tops, before the desiccation of such areas leaves them a bare white flat. In many parts of the Upper Basin, the crust or surface of the sinter flats can be broken through, exposing a structure that makes the origin of the deposit at once apparent to one familiar with the algous forms of hot water pools.

A number of simple experiments were made with the overflow of the Emerald Spring to determine, first, the rapidity with which the algæ establish themselves in new overflow channels, and secondly,

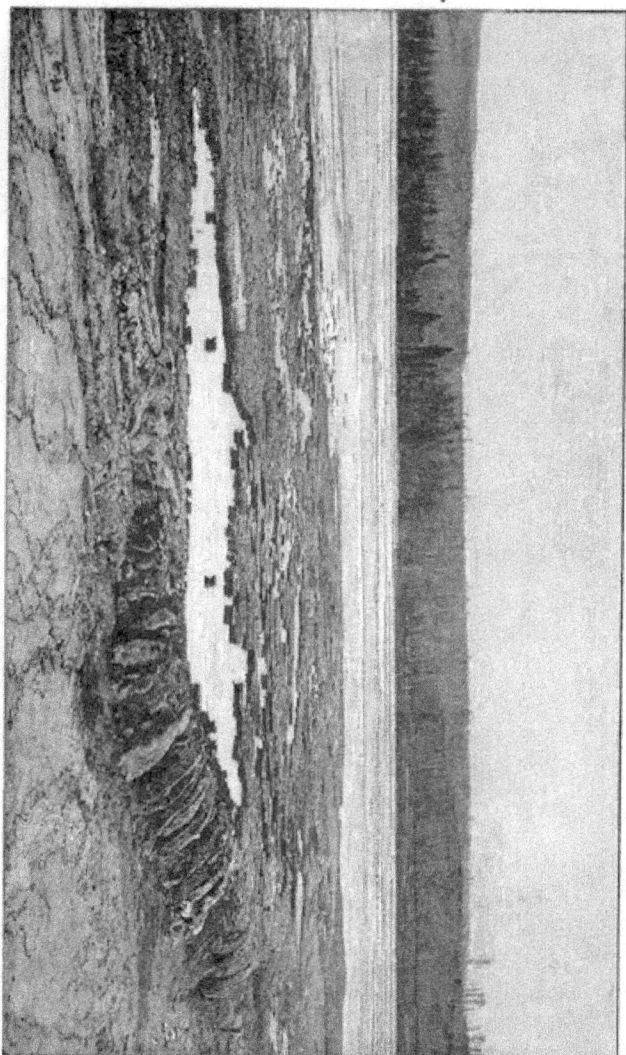

UPPER ALGÆ BASIN, JELLY SPRING, LOWER GEYSER BASIN.

the effect of a diminished supply and of temperature upon existing growths, and the final death of the algæ when the water is completely shut off. Cutting an outlet in the margin of the spring the outflow ran over a surface of compact, hard and dry sinter. On the second day this surface showed a very faint yellowish coloring; the third day this was easily noticed, and occurred in patches and not uniformly over the surface. Two weeks later, the greater part of the over-flowed area was covered with a fuzzy golden-green growth, which was coherent and membraneous in a few places, but which as yet showed no traces of the pillars and related forms found in the old basins. Where the rim of the spring had been cut, a shallow recess had permitted a partial cooling of the water, and an olive-colored, leathery sheet covered the floor. The current resulting from the overflow, immediately raised the temperature above this growth, which soon looked blistered and pale, and changed in the course of a few days to pale, yellowish green. Reducing the supply of the algæ channel and pools (Pl. LXXXII.) caused no change in the growth still covered by the water, in the first two days, but that portion of the growth left exposed to the sun soon began to dry and shrink. In twenty-four hours the dark emerald green of the leathery sheets had changed to dark purple, and where driest to black with a shining metallic luster. In drying, the siliceous jelly shrank considerably, and in consequence the surface layers had curled up in irregular patches, exposing the underlying layers of crimson jelly No odor was yet perceptible, but flies gathered thickly upon many parts of the decomposing vegetation. In those pools, where growth had pre-viously ceased, the algous forms were rapidly drying, the pink tint fading, and the more delicate parts already white and dry.

On the third day the surface layer of the leathery sheets was still more cracked up, the patches curled, with their edges white and dry, the underlying red jelly drying to rose pink, while the odor of decay-ing organic matter was strong and repulsive. The lowered temper-ature of the water had now effected the growth beneath it, and the olive and green flow showed patches of reddish brown, pink, and deep green. The supply of water being restored to portions of the water-way, the growth did not recover as rapidly as was expected. The lustrous black of the decaying vegetation changed in the course of a few days to spotted purple patches, but the red layers, still ex-posed, changed to salmon. There is no doubt that if the water sup-ply had not been restored the colors would have gradually faded out, leaving a white area of siliceous material as light as cork, where formed from the soft and jelly-like algæ, but heavier and denser, where the older forms had grown, this being the result at other places where similar pools are found.

If specimens of the different varieties of the growth be removed from the water and allowed to dry rapidly, the jelly contracts greatly in drying, the air-dried material being about one-third the bulk of the moist jelly. The gelatinous coating of the pillar and vase-shaped forms curls up in thin flakes, whose outer surface retains the color of the growth, exposing the light flesh-colored siliceous frame-work of the algæ.

The tendency of the algous growths to form terraced basins is beautifully illustrated in the basins supplied by the waters of the Jelly Springs at the base of the mound of the Fountain Geyser. In these basins the different stages of sinter forming are sharply drawn, from the soft and brightly colored jelly to a hard and stony sinter.

Pl. LXXXIV shows the uppermost of these basins; the dam ponding back the water is about a foot high, and is formed of a fibrous sinter, hard and stony below, but grading into a softer material of cheesy consistency above, passing into red and green algous jelly. The algæ of this pool or basin are brightly colored, and the forms resemble those of the Emerald Spring, but the pillars are taller, owing to the greater depth of the water.

In a lower basin, shown in Pl. LXXXV, the water is nearly cold, and though the forms are the same as those found in the basin above there is no trace of the red, yellow, and green algous jelly. A close view of the forms found in this basin is given in the cut (Fig. 56). In the basin, while covered by water, these peculiar structures are light pink, but they become white upon drying. The tops of the forms shown in the illustration are margined and capped by a very thin film of silica left by evaporation, and the small share which that agent takes in the formation of these deposits is shown in the relative proportion of this edging to the mass of pillars.

Pl. LXXXVII shows two of the forms from the basin figured in Pl. LXXXV. Fig. 1 is one of the finger-like pillars, which do not reach to the surface of the water. The specimen is six inches high and an inch in least diameter. The pure white surface is lined by little knife-edge ridges and dotted with spiny points of silica, all hung with small patches or shreds of a delicate web or film of silica, the remains of the algous jelly that once covered the surface. A transverse section shows the specimen to consist of a central core of white siliceous layers in the form of very thin concentric sheets or cylinders, surrounded by a loose wrapping of similiar paper-like sheets. The outer surface is hard, but brittle and easily broken. Such finger-forms frequently occur in clusters, sometimes of very different heights, and several often coalesce as they grow upward, and produce little pinnacled shapes. As already stated, in describing the algæ of the Emerald Spring, these pillars continue their upward growth when the algæ are living until their tops reach the water level, when, if the plant growth continues, a spreading top is formed, upon which evaporation leaves thin films of pearly silica.

One of the smallest of these curious, stony yet vegetable forms, is

MIDDLE ALGÆ BASIN, JELLY SPRING, LOWER GEYSER BASIN.

shown in Fig. 2, Pl. LXXXVII. The specimen figured is eight inches high, and shows in its graceful curves the bending of the original gelatinous material before the current of the basin. The broader base of the specimen is made of smaller spiny forms growing together and united to the base of the pillar. Above the middle the column expands into a hoop-shaped mass, crowned by irregular bands of pearly sinter. This specimen is also lined by the little ridges so prominent in the first figure, though they are much less noticeable and scarcely show on some parts of the specimen. Such forms reveal quite clearly their algous origin, but the stony masses found in a lower and empty basin, shown in Pl. LXXXVI, are apparently quite different in nature, though formed by the incrustation of the shapes shown in Fig. 56. This basin is the lowest of the series, and if some

FIG. 56. Algæ forms, Lower Geyser Basin.

cause had not operated to produce the death of the algæ, and an incrustation of the structures, before the filling up of the basin with their siliceous stems, the basin would now form only a bench, indistinguishable from the rest of the sinter flat above it. Fig. 3, Pl. LXXXVII, shows a specimen taken from this basin; the transverse section proves it to consist of a central form similar to Fig. 2, Pl. LXXXVII, covered with a mossy coating of silica, three-fourths of an inch thick, which rounds off and hides the outlines of the incrusted pillar. This coating has a rough coral-like surface, with clustered knobs of silica, which a lens shows to consist of delicate spicules of glassy sinter. The deposit is firm and hard, and the aggregated masses form a compact and solid sinter.

In the pools supplied by the Black Sand Spring, which are collectively known as Specimen Lake, the algæ are exactly like those described, save that they are generally slimmer and taller, often twelve or fifteen inches in length, and their tops, uniting, form a solid roof, often in turn the floor of a new basin, with a new growth of algæ. The pillars rarely grow solidly and closely together, so that specimens of the sinter are coral-like, the pillars coated with an efflorescent granular coating of silica. The desiccation of such areas leaves a deposit of sinter whose surface shows no trace of its origin and of the beautiful forms beneath, and such deposits occur in many places about the Geyser Basin.

The exact manner in which the algæ of these waters eliminate the silica from solution is not known, but the process appears to be due to the vital growth of the plant, for both the algæ filaments and their slimy envelope are formed of gelatinous silica. Upon the death of the algæ which have separated this jelly from the spring waters, there is a loss of a large part of its water, and a change to a soft, cheesy, but more permanent form. This dehydration is carried still farther if the silica be removed from the water and dried, but if allowed to remain in the cold water pools there is a further separation of silica, possibly due to organic acids, formed by the decaying vegetation reacting upon the silica salts of the water; this hardens the existing structures, in certain cases, and generally covers the pillars with a frost-like coating of silica.

In general, it may be stated that the large vase and pillar forms found in the algæ pools can be produced only by a concurrent life and death of these plants, the outer layers continually growing, the innermost dying. This is readily seen to be the cause of the peculiar structure of these forms. The central core is a pillar, sometimes hollow, sometimes solid, consisting of exceedingly thin superimposed layers of silica, each of which corresponds to a layer of algæ jelly, which has become hardened by the death of the plants and the loss of water. The column increases in diameter by the growth of the algæ at the surface, and a simultaneous death and hardening of the

inner layer of jelly. The algous envelope consists of two, three, or more thin membraneous layers, the outer, green, the inner, tomato red, these layers corresponding to the laminæ of the hardened inner core. The slimy, leathery sheets, so common in the cooler springs (100° F. to 135° F.), are similar in nature, and when dried are thin crusts of light, corky sinter. Another form, abundant about the Solitary Spring, where it has built up a sinter mound of considerable magnitude, consists of cushion-like masses of jelly, sometimes six inches thick, which, if removed and dried, shrivel up to less than half that thickness, and are exceedingly light and porous, floating on water. The under layer of such thick masses is decaying and changing to sinter, into which it can be traced in situ.

FIBROUS VARIETIES OF ALGOUS SINTER.

Besides the varieties of sinter formed by these vegetable jellies, there are two kinds of fibrous sinter, very abundant about some of the hot springs, and constituting an important part of the sinter deposits. The first, forming in the overflow channels of many of the geysers of the Upper Basin, is finely fibrous, consisting of layers one-sixteenth of an inch to half an inch thick, each stratum resembling a very fine thick white fur. This sinter is formed by the growth of the little algæ—*Calothrix gypsophila* Kg.—or the young form, *Mastigonema thermale*, the latter olive-colored and forming the sinter alluded to later in the section of the sinter walls of the crater of the Excelsior Geyser. The second form is fibrous, and occurs in rough, straw-like masses, with thatched arrangement. A coarse variety is due to a bright red species of algæ—*Leptothrix*—a finer variety to *Leptothrix* (or *Hypheothrix*) *laminosa*, a species found from 135° to 185° F., and ranging in color from white to flesh, pink, yellow, and red to green, as the water cools. The specimens determined came from the mounds of Sentinel Creek.

The proportion of algous sinter forming the deposits about the Geyser Basins is strikingly shown in the following section of the strata forming the wall of the Excelsior crater

	Inches.
21. Uppermost layer, fibrous, "furry" sinter	15
20. Cemented, sinter fragments	0.5
19. Fibrous sinter, brownish colored	3
18. Thatch-like, fibrous sinter	0.5
17. Cemented fragments	3
16. Thatch-like sinter	3
15. Fibrous	2
14. Cemented fragments	1
13. Thatch-like	12
12. Same, mixed with cemented material of same nature	2
11. Fibrous, 9 layers ⅓ inch to 1 inch thick	6
10. Cemented fragments, partly of organic origin	6
9. Fibrous	2

	Inches.
8. Flaky sinter formed by algous sheets	4
7. Fibrous and thatch-like, about equally divided	36
6. Fibrous	2
5. Flaky, pearly, algous	3
4. Thatch-like, brown	10
3. Fibrous, 10 to 20 layers	8
2. Cemented material	8
1. Fibrous	12

11 ft., 6 ins.

In this section fifty per cent. consists of the fibrous sinter formed by *Mastigonema*, 36 per cent. (4 feet, 2 inches) of the thatch-like or flaky sinter formed by the membranous algæ, *Leptothrix* ——.

The crater wall nearest the Prismatic Spring is 15 feet high, and the sinter may be thicker, as the underlying material is not exposed. This sinter, which forms a plateau covering many acres, has been formed by the vegetation nourished by the overflow from the Prismatic Spring, and the older layers have a terraced surface exactly like that of the deposit now forming about this spring.

RATE OF DEPOSITION OF SILICEOUS SINTER.

The pearl-beaded, coralloid forms of sinter found about spouting vents are formed very slowly. In one case, where the signatures of a party who visited the geysers in 1879 are known to be authentic, the pencil marks are covered by a glaze of silica but $\frac{1}{125}$ of an inch thick, or an increase of $\frac{1}{1000}$ of an inch a year, and this where the conditions for the formation of sinter by evaporation are quite favorable.

The difference between the rate of deposition of geyserite by these waters and those of the Norris Basin, notably by the water of the Opal Springs Coral, is shown by the fact that at the Opal Spring an incrustation of one-quarter of an inch formed in three weeks.

Other names, written upon the salmon-colored channels running into the Firehole, near the Castle and Saw-Mill geysers, show a growth of $\frac{1}{210}$ of an inch to $\frac{1}{8}$ of an inch a year, but this rate is effected by the combination of very favorable conditions for evaporation and the presence of algæ.

The fibrous sinters forming the flow of the geyser channel are composed of layers from $\frac{1}{30}$ to $\frac{1}{15}$ of an inch thick and averaging $\frac{1}{20}$ of an inch. If they represent a year's growth, and the evidence favors that view, the line of glassy silica separating them being formed during the winter, then the rate is $\frac{1}{20}$ of an inch a year.

On the other hand, the thick masses of jelly found in some of the overflow areas may form sinter with comparative rapidity. Thus the channel of the Beauty Spring, which contained no water in 1887, was filled with a growth of vegetable jelly 5 inches thick in 1888,

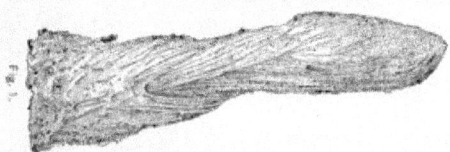

Fig. 1.

Fig. 2.

Fig. 3.

SINTER FORMS FROM ALGÆ BASINS.

nourished by the largely augmented overflow of the spring. A mass of this was cut out when the place was visited in July, and upon the sinter a new growth 1¼ to 1¼ inches thick had formed by October, seventy-three days' growth, while areas of what had been bright colored jelly in July had diverted the water by their growth, and were now hardened and pink, and rapidly passing into firm and solid sinter.

MICROSCOPIC EVIDENCE.

A microscopic examination of specimens collected at the Beryl Spring, Gibbon canyon, shows that the fibrous, asbestos-like material consists of minute tubes of glassy transparent silica corresponding to the filaments of the growing algæ. In this case the filaments appear to have been free from the enveloping jelly, which dries to an opaque white silica and hides the filaments and rods of most growths.

Thin sections of siliceous sinters fail to show the origin and nature of the deposit as clearly as had been hoped. A section of dried algous jelly from the Emerald Spring shows innumerable interlaced and interwoven filaments, with some glassy silica between. A hard fibrous sinter, formed by the long filamentary growth of an overflow channel, shows only traces of the algæ filaments under the microscope, but consists very largely of minute globules of glassy silica, varying somewhat in size and corresponding to those forming the cells of the algæ. These are held together in a cementing matrix of glassy amorphous silica. Thin sections of a sinter formed of broken fragments of algæ pillars, cemented into a firm hard sinter, shows a similar structure.

If many of the algous sinters fail to reveal an organic structure beneath the microscope, they are nevertheless easily distinguished from the more glassy and pearly sinters formed by evaporation. A thin section of a sinter from the Solitary Spring, Upper Basin, shows in marked contrast the numerous and extremely thin overlapping layers of lustrous pearl sinter formed by evaporation and the duller chalky white of the algous formation.

MOSS SINTER.

Besides the deposits of siliceous sinter formed by the algous vegetation of the hot waters, extensive deposits of sinter are found on the slopes below the Hillside Springs, which are due to the growth of mosses. These springs issue from the rhyolite slopes beneath the cliffs of the Madison Plateau, and the waters, whose temperature is 184° to 198° F., contain both silica and lime in solution, which they deposit in their downward flow. On the lower part of the slopes the water is cooled to blood heat, and has lost much of its lime and part of its silica. This part of the slope is terraced with basins suggesting those of the Mammoth Hot Springs, but covered with a bright green growth of moss. These basins are formed of a porous yellow sinter, full of moss stems, and often consisting entirely of these plant structures.

Chemical analysis shows this substance to be a true siliceous sinter (see analysis, p. 670). This sinter is not formed by evaporation, nor by any of the causes discussed in considering the precipitation of silica from solution, but it is due to the abstraction of silica from the water by the mosses covering the surface of the basins. This moss has been determined by Prof. Charles R. Barnes, of the University of Wisconsin, to be *Hypnum aduncum* var. *grasilescens* Br. & Sch.

DIATOM BEDS.

Besides the elimination of silica from the hot-spring waters by the algous growths living in them and by the mosses of the cooled water, there is a further secretion of that substance by several species of diatoms which live in the tepid waters of the hot-spring marshes, and, though they do not form siliceous sinter, their remains accumulate as beds of diatom earth that are often of great thickness and width.

It is well known that the single-celled algæ, called diatoms, possess in a remarkable degree the power of separating silica from solution to form the beautifully marked siliceous armor of the plant. In the ocean waters this action is the more remarkable because of the exceedingly small proportion of silica found in solution in the water, and the almost incredible activity necessary on the part of the plant to secure an adequate supply. As the silica of such dilute waters is not separable by any known chemical process, its elimination must needs be credited to some vital process of the plant growth, and it is this action which gives to this low form of life its importance as a geological agent. As the *Diatomaceæ* exist under very diverse and extreme conditions of environment, occurring in nearly every country pond and stream, as well as the icy waters of Polar seas, the heated currents of the tropics, and even the almost boiling waters of hot springs, we are not surprised to find them existing also in the siliceous waters of the Yellowstone Springs, which, indeed, seem peculiarly adapted to the needs and growth of these little plants. Investigation shows, however, that while diatoms occur in the ponds of the hot-water algæ, whose occurrence has already been described in detail, yet they are only found in abundance in the cooled, tepid waters of the springs. In such waters they are exceedingly abundant, and form the ooze of which the marshes of the geyser basins are so largely composed.

A typical marsh of this character is found near the beautiful Emerald Springs of the Upper Geyser Basin. A large part of this marsh is covered with a sparse growth of rushes and brackish-water vegetation, which of course is gradually filling it up and converting the bog into a fairly firm, grass-grown meadow bottom. But the greater area is at present quite wet, and its treacherous ooze and apparently bottomless depths will be long remembered by those who have ever tried to cross the marsh.

The waters of this area have in times past encroached upon the neighboring patch of timber, killing the trees, whose bare gray trunks stand upright in the ooze or lie scattered about and half immersed beneath the surface. A subsequent partial recession of the water has left a bare white strip between bog and woods, on which vegetation has as yet a precarious foothold, and the gaunt, bare poles of the dead pines rise up from a barren, powdery, white soil, evidently a dried portion of the marsh mud. The semi-liquid ooze of which the marsh consists proved upon examination under the microscope to be composed of the beautiful siliceous tests of various species of these minute plants. Samples of this material, which Dr. Francis Wolle, of Bethlehem, Pa., has kindly examined for me, were found to contain the following species :

Denticula valida Ped.	Achranthese.
D. elegans.	Cocconema.
Navicula major and N. viridis.	Fragilaria.
Epithemia (three species).	Eurotia.

The first-named species, *Denticula valida*, formed the bulk of the specimen, and also of the white pulverulent material at the margin of the bog, which microscopic examination showed to be the dried remains of the same diatoms. Samples from many other marshes of this character were examined and found to be formed of the same species.

The extensive meadows of the Lower Geyser Basin, the Norris Basin, Geyser Creek, and many other places are underlaid by beds of diatom earth composed of these same species, and where the wagon-road crosses these areas the ditches made alongside the road for drainage exposed the beds, while square blocks of the dried diatom earth lie scattered about at the side of the road. These and similar meadows are many square miles in extent, and the diatom beds are often two to three, and sometimes six to seven feet thick. Not seldom the meadows and diatom marshes overlie ancient hot-spring areas, the sinter flats and even the hot spring mounds and cones being completely covered and hidden by the covering of diatom ooze.

NATURE OF SILICEOUS SINTER.

Siliceous sinter, the siliceous deposit of thermal waters, is a variety of opal, occurring as a grayish white, or brownish incrustation about hot springs. Slightly different varieties have been described under different names : a pearly lustered specimen from Santa Fiora, Italy, being called Fiorite by Thompson; a filamentary sinter from

St. Michael, Azores, described as Michaelite; while the Iceland and
New Zealand sinters are known as geyserite.

Siliceous sinter varies much in appearance and in structure, according to its manner of formation; it is sometimes earthy and crumbly,
often finely laminated and shaly or light and porous, occasionally
fibrous or even filamentous, and rarely compact and flinty. It is
generally opaque, though often possessing a vitreous luster, and
rarely translucent, and it grades into hydrophane and hyalite by
alteration. The botryoidal and coralloidal forms are generally found
only about the mouth of geysers and steam vents, and are usually
more compact and translucent than other varieties. The sinters produced by algæ are generally very light, with an open, porous, almost
cavernous structure, but this is often altered to hard opal sinter by
infiltering water. The following table shows the composition of a
number of Yellowstone sinters, with the analyses of Michaelite and
the Iceland and New Zealand geyserites added for comparison:

Analyses of Yellowstone sinters.

	I.	II.	III.	IV.	V.	VI.	VII.	VIII.	IX.	X.
SiO₂, silica	89.54	81.95	98.88	93.37	89.72	87.67	82.29	86.03	92.67	...
Al₂O₃, alumina	2.12	6.49	1.73	1.16	} 1.02	0.71	} 1.30	1.21	0.80	}
FeO, ferrous oxide ..	Trace.	Trace.	0.14	Trace.			Trace.			
CaO, lime............	1.71	9.56	0.25	0.29	2.01	0.40		0.45	0.14	...
MgO, magnesia	Trace.	0.15	0.07	0.05	Trace	...	......	0.40	0.05	...
Na₂O, soda	1.12	2.96	0.28	0.11	......	0.82	......	} 0.38	0.18	...
K₂O, potash	0.30	0.65	0.23	0.02	......	Trace	......		0.75	...
SO₃, sulph. acid	Trace.	0.16	0.20	0.31	Trace.	......	......			...
Cl, chlorine....... ..	Trace	Trace	......	......	......	......	......			...
NaCl, sodic chloride.	......	......	0.18	0.08	......	......	......			...
Organic matter	......	......	......	1.50	......	......	......			...
H₂O, water..........	5.13	7.50	3.37	4.17	......	10.40	16.35	11.52	5.45	...
Total	99.92	100.02	100.33	100.41	100.09	100.00	100.00	99.99	100.04	...

No. I, is a compact white sinter from the mound of Old Faithful.
No. II, is a grayish sinter from the margin of Splendid Geyser.
No. III is a light porous algous sinter from Solitary Spring.
No. IV is the analysis of a dried specimen of jelly from Emerald
Spring.
No. V is the moss-sinter from Asta Spring, Hillside **Group.**
These analyses were all made by Mr. J. E. Whitfield for the
Geological Survey of the Yellowstone National Park.
No. VI is the analysis of Michaelite from the Azores, by Webster.[1]
No. VII is a geyserite from the mound of the Great Geyser of Ice-
and,[2] by Damour.

[1] Webster, Am. Jour. Sci., 1st series, 1821, vol. 3.
[2] Bull. Soc. Geol. de France, 2d series, 1848, vol. 5, p. 160.

No. VIII is a hard white sinter from the White Terrace, New Zealand, Mayer.[1]

No. IX is a white sinter from Steamboat Springs, Nevada, analyzed by Woodward.[2]

Analysis No. I shows the composition of a sinter in which evaporation and algous growth both cause the separation of the silica. The sinter from the Splendid Geyser, whose analysis is given in No. II, is a deposit formed without the aid or presence of plant life, wholly by the evaporation of the geyser water. The interior of the specimen is composed of irregular crinkled laminæ of greenish gray sinter, some of it possessing a nacreous lustre; this is covered by light gray fibrous sinter, the fibers short and perpendicular to the surface, and resulting from the growth of little spicules. The gray color is due to the impurity present, the analysis showing a comparatively large amount of alumina and soda, with a low percentage of silica. This is probably due to muddy sediment contained in the geyser water, which is left with the silica upon evaporation. The sinter from the Solitary Spring, No. III, is white, opaque, and very light and porous. It is the sinter resulting from the desiccation of an area of algous jelly, such as that abundant about this spring, and was produced by natural causes. Analysis No. IV shows the composition of a specimen of the algous jelly found at the Emerald Spring, removed from the water and dried in the sun. The siliceous residue of this jelly is light pink in color, very buoyant, floating readily upon water, and somewhat hygroscopic. The air-dried material lost 2.7 per cent. of its weight when dried at $100°$ C.

Analysis No. V is of the straw-colored moss sinter from the basins of the Asta Spring. The structure of the moss is perfectly preserved, and much of the sinter is composed entirely of the moss stems; but other parts have the space between filled with friable white silica, with occasional botryoidal concretions of light gray opal.

The analyses show the greater purity of the sinters formed by algæ, such sinters having less alumina and alkalies, a lower percentage of water, and a correspondingly larger amount of silica. This greater purity is probably to be explained by the fact that the silica of such sinters has been extracted from the water by the vital growth of the plants, while sinters formed by evaporation contain a greater or less amount of kaolin, generally carried in minute quantity in suspension by the geyser waters. It is to such earthy impurities that we must ascribe the differences in the analyses of Iceland sinters made by different chemists.

The physical differences in the unaltered sinters formed by evapo-

[1] Peterm. Geog. Mitt., 1862, p. 266.
[2] Arnold Hague, Geol. Ex. 40th Par., 1877, vol. 2, p. 826.

ration and those of algous origin is generally quite marked, the former being translucent, or vitreous, hard, and heavy, while the algous sinter is opaque, white, and often chalk-like in appearance.

SILICEOUS SINTERS FROM NEW ZEALAND.

Through the courtesy of Prof. F. W. Clarke, a small collection of siliceous sinters from the hot springs of New Zealand, recently received by the United States National Museum, has been placed at my disposal for examination and comparison with the extensive series of sinters collected by the Yellowstone Park Survey from the hot springs and geysers of that region.

The New Zealand collection, though small, contains examples of many different varieties of this form of opal—the result of diverse conditions of deposition and occurrence.   Most of the specimens come from Rotorua, the sanitarium of New Zealand, situated on the south-west shore of the lake of the same name.   This locality was long known as Ohinemutu, the name of the Maori village, for the natives of New Zealand utilized the hot waters of these springs for cooking and bathing before the discovery of the islands by Captain Cook. The government, appreciating the therapeutic value of the waters, has leased the ground from the Maoris and erected extensive bathing pavilions and bath pools, where the different varieties of waters may be tried under the direction of a government physician.   The place was formerly the starting point for the famous terraces of Rotomahana (the "Warm Lake"), which were destroyed by the eruption of Mount Tarawera in 1886.   But Rotorua is itself interesting ; the lake is six miles across, with a picturesque island in its center, and surrounded by a chain of blue mountains, while the ponds and wells of hot water and the neighboring geysers of Whakarewarewa are only rivaled by those of the Yellowstone and the famous fountains of Iceland.

The hot springs vary in character from the clear and sparkling, albeit boiling, alkaline siliceous waters of Madame Rachel's Bath, supposed to renew beauty, if not youth, to chocolate-colored and ill-smelling sulphurous pools and strongly acid waters.   The following analyses, made by William Skey, the Government analyst, show the character of several types of these waters, the analyses being reduced from grains per gallon to parts per thousand.

Analyses of New Zealand spring waters.

	Madame Rachel's Bath, alkaline, 174° F.	Priests' Bath, strongly acid.	Hot Pool, intensely acid, 200° F.
SiO_2, silica, free	0.0838	0.2630	0.1957
Na_2SiO_3, sodium silicate	0.2601	....	.....
$CaSiO_3$, silicate of lime	0.0905	....	.....
$MgSiO_3$, silicate of magnesium ..	0.0155	....	.....
NaCl, sodium chloride	0.9918	....	0.5210
KCl, potassium chloride	0.0487	....	....
$NaSO_4$, sodium sulphate	0.1685	0.2748	0.2043
$CaCl_2$, chloride of lime	.. ..	....	0.1025
$MgCl_2$, chloride of magnesium ...	....	....	0.0148
Al_2Cl_6, chloride of aluminium ..	....	....	0.0617
$CaSO_4$, sulphate of lime......	.. ..	0.1698	....
$MgSO_4$, sulphate of magnesia ...	....	0.6432	....
$Al_2(SO_4)_3$, sulphate of aluminium.	......	0.0265	... .
$Fe_2SO_4)_3$, sulphate of iron.	......	0.1770	....
H_2SO_4, sulphuric acid	....	0.3160	.....
HCl, hydrochloric acid	....	0.0521	0.2915
Total....	1.6633	1.3821	1.3091
Free H_2S	....	0.6425	0.1255
Free CO_2	....	0.0308	0.0177

The alkaline water of Madame Rachel's Bath is said to deposit silica quite rapidly, "a ti-tree twig immersed in the water a week or two resembling a branch of coral,"[1] and the acid waters of the Hot Pool form mineral mushrooms of muddy sinter in the shallow parts of the spring.

It will be noticed that in these analyses the bases have been combined with silica by the analyst, who states that the conditions under which the waters occur are incompatible with the existence of carbonates, though the analyses of Yellowstone waters show that the fixed CO_2 of those waters must exist as carbonates.

The siliceous sinters from Rotorua vary from pulverulent deposits of impure silica to dense, white opal sinters. Two of the specimens were evidently formed about spouting vents, showing the peculiar structure and beaded surface produced by the evaporation of spattered drops of water. Such sinters, to which the name of *geyserite* may be most properly applied, are very common about the Yellowstone geysers, occurring often in beautiful coralloidal forms, sometimes possessing a bright pearly luster. The New Zealand specimens are parts of an old deposit formed in this way and consist of numerous little pillars formed of many convex layers of pink and white silica, resembling a pile of minute caps, one upon another. This geyserite is wholly the result of evaporation, which adds film after film of glassy silica to the surface of the deposit, as often as wet by the

---

[1] J. A. Froude, Oceana, p. 236.

steam or spray from the geyser. An analysis of this sinter is given
in the table following (p. 675). Specimens of what may be called incrustation sinter resemble a handful of hay crushed in the hand
and coated with white silica. The coarser stalks are hollow tubes
with rough and coral-like outer surface with the finer fibers forming
gnarled and botryoidal masses. Where the incrusting process has
been carried still further the thickened coverings of silica unite and
a compact sinter with but small cavities results. Such sinters are
also formed by evaporation, on the exposure of the water to the air.

Two of the specimens are of especial interest because their structure indicates that the algous life of the hot waters of Rotorua produced siliceous sinter. This action of hot water algæ has been
studied in the waters of the Yellowstone springs and it has been
shown that these plants abstract silica from the waters in their
growth, forming a mass of jelly which hardens upon the death of
the plants, when it is further incrusted by silica precipitated by
the decaying vegetable matter. In this way vast areas of sinter,
often many feet thick, have been formed in the Yellowstone Park.
The specimens from Rotorua show two distinct forms of algous sinter. The first is that produced by membranous sheets of red or
green algæ, resembling certain more familiar forms of seaweeds,
common not only in the cooler waters of the Yellowstone but in
warm waters all over the world, being described as "sheets of a
slimy confervoid growth "[1] in the Rotorua waters. This sinter is
creamy pink, showing a wavy and very thinly laminated structure
with occasional vesicular blisters lined with red and green patches
presumably the remains of algæ. It resembles so closely the sinters
formed by drying the algous jellies of the Yellowstone springs that
a similar mode of formation seems probable.

The second specimen is quite different in structure, consisting of
several layers of fibrous silica, the fibers all perpendicular to the
layers and resembling a very fine and short, thick, white fur. The
exact counterpart of this sinter occurs at many localities in the geyser basins of the Yellowstone, notably about the Prismatic Spring
and the overflow channels of Old Faithful. It forms over one-half
of the section of 15 feet of sinter exposed in the crater walls of the
Excelsior Geyser. This sinter we know to be the result of the
growth and incrustation of little algæ, which form a cedar-colored
(*Calothrix gypsophila* Kg., or olive (*Mastigonema thermale*) slippery
coating on the surface of the deposit. The analogy is so perfect
that there seems but little doubt that the New Zealand sinter is the
result of the growth of similar or allied algæ.

Other specimens of the hot-water deposits of the Rotorua Springs
resemble blocks of diatomaceous earth and vary from a loosely com-

[1] Skey Trans. N. Z. Inst., vol. 10, p. 433.

pacted mass of pulverulent silica to a dense and almost jaspery sinter. The impalpable particles composing this material are angular and consist almost wholly of milky or transparent glassy silica. Analysis No. III of the following table shows this substance to be a mixture of clay and silica of the same composition as the material incrusting logs immersed in the siliceous waters of the Yellowstone, and often lining the hot-spring bowls.

The following analyses of three types of the Rotorua sinters were made by Mr. J. Edward Whitfield:

Analyses of Rotorua Sinters.

	I.—Geyserite.	II.—Algæ sinter.	III.—Pulverulent silica.
Si O₂, silica	90.28	92.47	74.63
Al₂ O₃ (+Fe₂ O₃)..........	3.00	2.54	15.20
Ca O, lime	0.44	0.79	1.00
Mg O, magnesia	Trace.	0.15	Trace.
Na₂ O, soda.............	....	....	0.30
K₂ O, potash	....	....	1.02
Ignition	6.24	3.99	7.43
Total.............	99.96	99.94	99.07

The remaining specimens from Rotorua consist of a hot spring sandstone, produced by the cementation of particles of rhyolitic glass, feldspar, and quartz by the siliceous waters of the springs. The fragments are uniformly angular, well assorted in layers, and bound together by transparent or milky silica. The latter usually forms a very small part of the bulk of the specimen; in only one case does it sheath the grains in a coating of silica, and form a noticeable part of the deposit. Under the microscope thin sections show no traces of enlargement of the crystal fragments. Though a true hot spring deposit, this material can not claim the name of siliceous sinter. Two of the specimens contain very showy leaf impressions, but the details of veination are not preserved, and the woody tissue is absent. A white "mineral wool" occurring in other specimens of this nature is perfectly silicified woody fibre. Such sandstones are common about the hot springs of the Norris and Shoshone Geyser Basins and other localities of the Yellowstone Park, where they are formed by the cementation of material washed into the springs from the surrounding slopes of disintegrating or decomposed rock.

Besides the sinters from Rotorua, the collection contains a few specimens from the famous White Terrace. This sinter opal closely resembles the deposits of the Coral Spring, whose water, like that of the lost White Terrace, is opalescent with silica, which is carried in pseudo-suspension, and which rapidly coats articles immersed in it. Two jaspery sinters from Whangerei—a seaport about eighty miles

north of Auckland—are very beautifully colored, in red and green
bands, but are of no especial interest.

No information is obtainable relative to the comparative abundance
of the different types of sinter, but the prevalence of acid and comparative scarcity of alkaline waters shown by the list of springs published by Dr. Hector leads to the belief that algous sinter forms a
smaller proportion of the siliceous deposits than it does at most of
the geyser basins of the Yellowstone, where the waters are chiefly
alkaline. The general character of the springs shows that Rotorua
resembles the Norris Basin more closely than any other locality in
the Park.

SUMMARY.

In the light of the knowledge gained in the Yellowstone Geyser
Basins, the observations of Prof. W. H. Brewer, referred to in the
first part of this paper, acquire a new interest, and it seems quite
probable that the gelatinous silica containing algæ which he found
at Steamboat Springs, Nevada, may resemble that so abundant in
the hot waters of the Park, and that a part, at least, of the siliceous
deposits found at Steamboat Springs may have been formed by
algous life. The fibrous sinter from the Azores, called Michaelite,
occurring about springs where algous vegetation was found to be
abundant, certainly suggests a possible like origin. The data accessible are far too meager to hazard any conjectures as to the nature
of the Iceland or New Zealand sinters, but the occurrence of algæ in
these waters is significant in this connection.

It is believed that the facts recorded in the preceding pages establish—

1. That the plant life of the calcareous Mammoth Hot Springs
waters causes the deposition of travertine, and is a very important
agent in the formation of such deposits.

2. The vegetation of the hot alkaline waters of the Geyser Basins
eliminates silica from the water by its vital growth and produces
deposits of siliceous sinter.

3. The thickness and extent of the deposits produced by the plant
life of thermal waters establishes the importance of such vegetation
as a geological agent.

www.ingramcontent.com/pod-product-compliance
Lightning Source LLC
Chambersburg PA
CBHW021956190326
41519CB00009B/1283